Scientific Fundamentals of Robotics 4

M. Vukobratović
N. Kirćanski

Real-Time Dynamics of Manipulation Robots

With 43 Figures

Springer-Verlag
Berlin Heidelberg New York Tokyo

D. Sc., Ph. D. MIOMIR VUKOBRATOVIĆ, corr. member of
Serbian Academy of Sciences and Arts
Institute »Mihailo Pupin«, Beograd,
Volgina 15, POB 15, Yugoslavia

Ph. D. NENAD KIRĆANSKI
Institute »Mihailo Pupin«, Beograd,
Volgina 15, POB 15, Yugoslavia

ISBN 3-540-13072-1 Springer-Verlag Berlin Heidelberg New York Tokyo
ISBN 0-387-13072-1 Springer-Verlag New York Heidelberg Berlin Tokyo

Library of Congress Cataloging in Publication Data.
Vukobratović, Miomir.
Real-time dynamics of manipulation robots.
(Scientific fundamentals of robotics ; 4)
(Communications and control engineering series)
Bibliography: p.
Includes index.
1. Robots--Mathematical models.
2. Digital computer simulation.
3. Real-time data processing.
  I. Kirćanski, N. (Nenad)
  II. Title.
  III. Series.
  IV. Series: Communications and control engineering series.
TJ211.V85   1985   629.8'92   85-2717

Offsetprinting: Mercedes-Druck, Berlin
Bookbinding: Lüderitz & Bauer, Berlin
2061/3020 5 4 3 2 1 0

About the Series:
»Scientific Fundamentals of Robotics«

The age of robotics is the present age. The study of robotics requires different kinds of knowledge multidisciplinary in nature, which go together to make robotics a specific scientific discipline. In particular, manipulator and robot systems possess several specific qualities in both a mechanical and a control sense. In the mechanical sense, a feature specific to manipulation robots is that all the degrees of freedom are "active", i.e., powered by their own actuators, in contrast to conventional mechanisms in which motion is produced primarily by the so-called kinematic degrees of freedom. Another specific quality of such mechanisms is their variable structure, ranging from open to closed configurations, from one to some other kind of boundary conditions. A further feature specific of spatial mechanisms is redundancy reflected in an excess of the degrees of freedom for producing certain functional movements of robots and manipulators.

From a control viewpoint, robot and manipulator systems represent redundant, multivariable, essentially nonlinear automatic control systems. A manipulation robot is also an example of a dynamically coupled system, and the control task itself is a dynamic task.

The basic motivation for establishing the conception of this series has consisted in an intention to clearly define the role of dynamics and dynamic control of this class of system. The associates who have been engaged in the work on this series have primarily based their contributions on the development of mathematical models of dynamics of these mechanisms. They have thus created a solid background for systematic studies of robot and manipulator dynamics as well as for the synthesis of optimal characteristics of these mechanisms from the point of view of their dynamic performances. Having in mind the characteristics of robotic systems, the results concerning the problems of control of manipulation robots represent one of the central contributions of this series. In trying to bridge, or at least reduce, the gap existing between theoretical robotics and its practical application, considerable efforts have been made towards synthesizing such algorithms as would be suitable for implementation and, at the same time, base them on sufficiently accurate models of system dynamics.

The main idea underlying the conception of the series will be realized: to begin with books which should provide a broad education for engineers and "create" specialists in robotics and reach texts which open up various possibilities for the practical design of manipulation mechanisms and the synthesis of control algorithms based on dynamic models, by applying today's microelectronics and computer technologies.

Those who have initiated the publication of this series believe they will thus create a sound background for systematic work in the research and application of robotics in a wider sense.

Belgrade, Yugoslavia, February 1982   M.Vukobratović

Preface

This is the fourth book from the Series "Scientific Fundamentals of Robotics". The first two volumes have established a background for studying the dynamics and control of robots. While the first book was exclusively devoted to the dynamics of active spatial mechanisms, the second treated the problems of the dynamic control of manipulation robots.

In contrast to the first two books, where recursive computer-aided methods for setting robot dynamic equations where described, this monograph presents a new approach to the formation of robot dynamics. The goal is to achieve the real-time model computation using up-to-date microcomputers. The presented concept could be called a numeric-symbolic, or analytic, approach to robot modelling. It will be shown that the generation of analytical robot model may give new excellent possibilities concerning real-time applications. It is of essential importance in synthesizing the algorithms for nonadaptive and adaptive control of manipulation robots.

If should be pointed out that the high computational efficiency has been achieved by off-line computer-aided preparation of robot equations. The parameters of a specified robot must be given in advance. This, after each significant variation in robot structure (geometrical and dynamical parameters) we must repeat the off-line stage. Thus is why the numerical procedures will always have their place in studying the dynamic properties of robotic systems.

This monograph is organized in 5 chapters.

Chapter 1 provides a review of computer-aided methods for mathematical modelling of open mechanical configurations of robotic systems, which have a tendency towards real time. A series of methods is presented, among them those by Hollerbach, Renaud, Kahn, Kane, and other authors, which belong to the class of numerical methods and are based on either Lagrange's second-order equations or Newton-Euler's or Appel's equations.

Chapter 2 presents a new formulation of motion equations of open-loop
mechanisms in a closed form. Such models are shown to have a very compact form with a clear insight into geometrical, kinematic and dynamic
robot parameters. In addition, this chapter presents the algorithms for
constructing closed-form linearized models, significant for robot stability testing, and for control synthesis under small perturbations
acting on the system around nominal trajectories.

Chapter 3 describes new concepts for numeric-symbolic robot modelling.
Relevant notions and definitions are introduced, numeric-symbolic representation of system variables is given, and the notion of unified
matrix ("polynomial matrix") for describing arbitrary system variables
is introduced.

Chapter 3 contains the complete algebra of polynomial matrices. Special
attention is devoted to the problem of optimizing the polynomial matrices.
These results have been used to form the algorithm for generating numeric-symbolic robot models. These results have also allowed the construction of linearized, sensitivity and approximate models.

Chapter 4 describes the procedure for the computational optimization of
analytical model and for the generation of program code for real-time
operation. The algorithm for calculating the polynomial matrices with
a minimum number of numerical operations is elaborated. This chapter
also presents the very modern approach to problem solving using an "expert program". The task of this program is to "write" the robot model
in a desired programming language.

In Chapter 5 all previous results are illustrated by examples of actual
robots which have been implemented in industrial technological processes. The robots have been chosen according to the structure of their
mechanisms. One of them is of anthropomorphic structure, the second of
cylindrical, and the third of arthropoid configuration. Such structures
have been chosen to illustrate the dependence of the complexity of mathematical models on the complexity of robot structure. The procedure
for computational optimization of numeric-symbolic models is illustrated at the end of each example. This includes illustration of all notions introduced in Chapter 4, such as the sequence of optimal calculation, elementary graph, matrix representation of graph, etc. Special
attention is given to program code generation using the expert-program.
From the obtained program codes, the exact number of floating-point

multiplications and additions is determined. This is very important for microcomputer implementation.

Similarly to the previous two volumes of the Series, the present book is also intended for students enrolled in postgraduate robotics courses and for engineers engaged in applied robotics, especially in studying dynamics of robotic systems, synthesizing control algorithms and, first of all, in their microcomputer implementation for actual, complex tasks arising in industrial robotics. The authors also think that this, essentially monographic, work may, with no great difficulties, be used as teaching material for subjects from the field of robotics at graduate studies the organization of which is in progress.

The authors are grateful to Miss G. Aleksić for her help in preparing English version of this book. Our special appreciation goes to Miss V. Ćosić for her careful and excellent typing of the whole text.

October 1984
Belgrade, Yugoslavia   The Authors

Contents

Introduction

The application of control theory to large-scale mechanical systems, such as robots, aircrafts, complex hydraulic systems, etc., represents an extremely difficult problem because of the prominent nonlinearity and complexity of mathematical models of these systems. With industrial robots, which will be treated in this book, the application of such theory and the development of new control algorithms are unavoidable in order to achieve high positioning accuracy and speed. Several control concepts employing partially simplified or even exact mathematical models in on-line operation have so far been brought to an advanced level. In on-line control, the calculation of model equations must be repeated very frequently, preferably at sampling frequency which is not lower than 50 Hz, since the resonance frequency of most mechanical manipulators is about 10 Hz. However, the problem of solving the dynamic equations of manipulators in real time, i.e. after each 20 ms or less, by means of today's microcomputers is rather difficult and complex.

In explaining a new approach to solve this problem, which we shall follow in this book, let us first mention some results in robot modelling achieved so far. It is possible to systematize these results according to different criteria. Thus, methods may be classified with respect to the laws of mechanics on the basis of which motion equations are formed. Taking this as a criterion, one may distinguish methods based on Lagrange's, Newton-Euler's, Appel's and other equations. The property of whether the method yields the solution to both the direct and the inverse problem of dynamics or not, may represent another criterion. The direct problem of dynamics refers to determining the motion of the robot for known driving forces (torques), and the inverse problem of dynamics refers to determining the driving forces, given a manipulator motion. Clearly, the methods which yield the solution to both problems of dynamics are of particular importance. The number of floating-point multiplications/additions required to form the model is yet another criterion for comparing the methods. This criterion is certainly the most important one from the point of view of their on-line applicability.

The first result in the class of Lagrange's methods, which are used to solve either the direct or the inverse problem of dynamics, was reported by Uicker. Since the method referred mostly to certain classes of closed-loop mechanisms, Kahn elaborated an algorithm for modelling open kinematic chains. The method was modified and preformulated by Woo and Freundenstein through introducing screw-calculus. Improved versions of this method in the sense of a reduced number of numerical operations were developed by Mahil and Renaud. A property common to these algorithms is that they do not employ recursive relations but closed-form expressions for the elements of model matrices. However, the number of multiplications/additions in these methods depended on n^4 (n - the number of degrees of freedom of robot. Thus, several thousand multiplications and additions were required for manipulators with three degrees of freedom. The number of operations for manipulators with six degrees of freedom was more than ten times larger. Thus, it is difficult to apply these algorithms on contemporary microcomputers in real time. With up-to-date 16-bit microcomputers (with arithmetic coprocessors or appropriate hardware support), one multiplications and one addition take about 0.1 ms (further in this book, such a microcomputer will be termed as reference microcomputer). This, in turn, means that about two hundred floating-point multiplications and additions are implementable in 20 ms, and this is considerably less than the number of operations required in the above-mentioned methods.

Vukobratović and Potkonjak developed an algorithm which employs recursive expressions for the construction of matrices of dynamic models. Although the number of operations was thus reduced by a number of times in comparison with Kahn-Uicker's method, even this method could not be easily implemented in real time, since it required more than two thousands of floating-point operations for a standard manipulator configuration.

As may be seen from the above discussion, the algorithms based on Lagrange's equations, which yield the solution to both the direct and the inverse dynamic problem, require too many computer operations to be applicable in real-time. This is why Waters and Hollerbach developed algorithms for solving only the inverse problem of dynamics on the basis of Uicker-Kahn's method. These procedures allow the determination of driving forces (torques), for given a manipulator motion. The number of floating-point multiplications was thus reduced to dependence on n^2, or even n. These savings result from the fact that the inertia matrix of manipulator model is not explicitly calculated, i.e. the second derivatives of joint coordinates implicitly figure in dynamic equations. Therefore, such algorithms are not applicable to all control concepts. In addition, regardless of the dependence of the number of multiplications on n, these methods are also hardly implementable in real time.

Application of the Newton-Euler dynamic equations to modelling of joint--interconnected rigid bodies may be found in papers by Kane. Kane derived dynamic equations not restricting his attention to robotic systems, but for a somewhat broader class including space vehicles (satellites) with several joint-interconnected rigid bodies. Huston made a concretization of Kane's results for robotic systems and elaborated its program implementation. These algorithms solve both the direct and the inverse problem of dynamics, using closed-form expressions for the elements of dynamic model matrices. However, similarly to Kahn-Uicker's method, the number of operations required to form the model by these algorithms is very large. A more efficient computer method was elaborated by Vukobratović and Stepanenko through introducing recursive relations into the solution of either the direct or the inverse problem of dynamics. It was shown that the number of floating-point multiplications needed in this method was proportional to n^3. Walker and Orin gave a new organization of the calculation and thus reduced the number of operations to dependence on n^2. However, the number of operations was not sufficiently reduced to make the method implementable in real time. A special case on Newton-Euler method which yields the solution to the inverse problem of dynamics was treated by Luh, Walker and Paul. By expressing the equations of motion of each link exclusively in local coordinate systems, they obtained a procedure much more efficient than preceding ones. According to this algorithm, the number of floating--point multiplications is reduced to dependence on n. Thus about 400 operations are needed for manipulators with 3 degrees of freedom. Although considerably smaller than the previously mentioned, even this number is nearly twice larger than the number of operations which could be achieved in 20 ms on our reference microcomputer. In addition, this algorithm is not applicable to all control concepts, since it does not give the inertia matrix of the system but only driving forces and torques. To implement real-time control Luh and Lin developed, employing the preceding algorithm, a procedure for the calculation of driving forces by parallel calculation on a computer with several central processor units. The procedure was illustrated by the example of six-

4

-degree-of-freedom Stanford manipulator controlled by six processor
units. It was shown that the calculation of the forces (torques) of actuators required about 20 floating-point multiplications and additions
per one processor unit.

We see that real-time control of manipulators with no more than 3 degrees of freedom is hardly implementable by using a single processor
unit. The following essential question thus comes up. Is the number of
floating-point operations in previous methods minimal? Before answering
the question, let us note a property common to all mentioned methods.
The described methods are independent on manipulator mechanical design
and general laws of kinematics and dynamics for a system of rigid bodies, are emebedded into the procedures for dynamic modelling. One can
recognize that, given a manipulator, the actual analytical model com-
prises a considerably less number of numerical operations. Certain in-
dications about the validity of this hypothesis were given by Aldon,
the authors of this book and Renaud. For example, it was shown that the
construction of the model of anthropomorphic manipulator with three degrees of freedom required no more than 44 floating-point multiplications and 23 additions, and this is over 10 times less than in any of
the above-mentioned methods. Only 352 multiplications and as many additions were shown to be required for a manipulator with five degrees
of freedom. However, none of the mentioned results gives a solution to
the problem of computer-aided generation of an analytical model for an
arbitrary, given manipulator. This problem will be considered in detail
in this book.

A new computer-oriented method for dynamic robot model generation, based
on closed-form dynamic equations, will be presented in Ch. 2. The algorithm is nonrecursive and convenient for computer implementation due
to its compactness and explicit dependancy on physical parameters.
Then, several theorems for linearized and sensitivity closed-form model
construction are proved.

Using the developed dynamic equations, several dynamic model features
are mathematically demonstrated. Some of them are well-known, such as
simmetry and positive definitness of inertial matrix. There are also
several features, which have not been published in robotic books so far.
For example, antisymmetry of dynamic matrices describing Coriolis' and
centrifugal effects will be proved.

The general representation of variables which figure in equations of

robot dynamics is introduced. This representation is called "numeric-
-symbolic", because the parameters are treated numerically, while variables' dependancy on joint coordinates is treated symbolically. The
unifoldness of such representation is proved. It is also shown that the
variables which describe system dynamics can be uniquely represented in
a polynomial form. Further, it is shown that these polynomials may be
represented in a matrix form. As these matrices describe the structure
of the polynomials these matrices will be called "polynomial matrices".
Some fundamental features of the matrices are stated by several theorems and proved. These results are important because they estimate the
computer memory needed for storing various numeric-symbolic expressions,
i.e. matrices.

In the text to follow, algebra of polynomial matrices is developed. The
algebra is based on specially introduced polynomials which are shown to
be equivalent to numeric-symbolic expressions. After that, the set of
polynomial matrices of variables of a robotic system is shown to con-
stitute a vector space. For this purpose, the operations such as addition, multiplication by a real constant as well as scalar and vector
multiplication are defined. Neutral and inverse elements with respect
to a given variable are also defined. Further, the introduced algebra
is shown to be very suitable for digital computer program implementation.

The basic problem, considered in the next part of the book, is related
to the uniqueness of these polynomials. It is shown that for a given
model variable there exists indefinite number of different polynomial
matrices which correspond to it. This non-uniqueness is essentialy
caused by the nonuniqueness of trigonometric relations determining the
dynamic variables. Now, we encounter the problem of determination of a
unique representative for a set of equivalent matrices. It is shown
that such a representative exists and corresponds to the matrix with
minimal number of rows. An efficient algorithm for dimension minimization suitable for computer implementation is developed and illustrated
on several examples.

The generation of the complete nonlinear dynamic robot model in analytic form, based on the described theoretical results is presented in
the sequel. Using the obtained analytic expressions, the linearized
robot model in numeric-symbolic domain is also developed. The linearized model is obtained by partial derivation of polynomials. This operation is formalized within a theorem, and shown to be easily imple-

mentable on computers. A sensitivity model and an approximate analytical model are also developed.

The polynomials mentioned above are multivariable. Now we can state one of the most important issues concerned with the implementation of the previous theory: how to calculate the polinomials with minimal number of floating-point operations. This problem is considered in this book in details. A procedure for optimal (eventually suboptimal) sequence of operations for computing the polynomials is given. Within the scope of such a procedure several notions, cush as degree of an ordered set, inclusion between ordered sets etc., are introduced. The procedure is composed of several steps by which the global minimization is replaced by a series of independent successive minimizations. In the general case such an approach might not bring to an absolute minimum of operations. On the other hand, this extraordinary simplifies the algorithm making it attractive for implementation.

Previous results provided the basis of constructing an "expert-program" which generates a program for calculating the analytical robot model in a desired computer language. This program code can now be implemented on microcomputers belonging to control systems of the robot for which the model was formed.

For the sake of recapitulation, it should be pointed out that the numeric-symbolic modelling differs from previously developed numerical algorithms in several basic ideas:

- the analytical structure of dynamic robot equations is precalculated off-line after the series of mathematical optimizations,

- numeric-symbolic algorithm is not composed of different kinematic and dynamic laws. On the contrary, it contains very compact analytical forms which are optimally condensed in numerical sence.

Finally, one should consider prospectives for the further development of both analytical and numerical methods for modelling the dynamics of robotic mechanisms.

As far as the analytical method is concerned, the possibility for increasing its efficiency should be sought in developing the algorithms for global minimization of polynomials. This would be particularly important for manipulation robots having six and more degrees of freedom which are becoming more and more interesting for application. It should be stated that, for manipulators with no more than five degrees of freedom, successive minimization of polynomials yields results acceptable for practical application and microcomputer implementation.

A further reduction of numerical complexity in the recursive methods is also achievable. One illustration of this statement is the already obtained result which is presented in this book, and is described in detail in [68]. It refers to a new, and the most efficient so far, numerical computer algorithm which gives, for manipulation robots with six and more degrees of freedom, a smaller number of numerical operations than that achievable by using the analytical method which is at present based on the successive minimization of polynomials. However, the elaboration of global optimization of polynomials will offer more efficient analytical procedure for forming robot dynamic models with a minimal number of floating-point multiplications/additions.

This is of great interest and importance for nonadaptive and especially adaptive, control of manipulation robots.

Chapter 1
Survey of Computer-Aided Robot Modelling Methods

1.1. Introduction

The past fifteen years have witnessed a rapid development of robot modelling methods. These methods are significant in many respects. For example, they play a decisive role in the problems of robot design, control synthesis, and real-time control itself. It is possible to systematize these methods according to different criteria. Most frequently, the systematization is performed according to the applied laws of mechanics. Three basic groups of methods are distinguishable according to this criterion:

1. methods based on Lagrange's equations,

2. methods based on Newton-Euler's dynamic equations, and

3. methods based on Appel's equations.

The most important algorithms from these three groups will be presented in this chapter. The following essential characteristics will be stated for each of these algorithms:

1. whether they enable the solution of the direct and inverse problems of dynamics,

2. whether they employ recursive relations or closed-form expressions only, and

3. the number of numerical operations required for model computation.

The direct problem of dynamics refers to determining robot motion, given a driving forces, while the inverse problem refers to determining the driving forces for given motion. Clearly, the methods enabling the solution to both the problems are of particular importance. The number of numerical operations required for forming a model is an especially important criterion for comparing these methods. This number establishes whether or not a method is applicable to real-time dynamic control of robots.

Further, we distinguish numerical and symbolical methods. The numerical methods were almost exclusively developed during the past ten years. In these methods, a numerical quantity is associated to each variable which is included in model construction. On the other hand, in symbolic methods variables represent collections of symbols which describe their dependence on system parameters and states (position and velocity). The advantages and disadvantages of both groups of methods will be presented in detail in this chapter. It will be shown that the numerical methods are hardly implementable in real time because of their complexity and the large number of numerical operations. On the contrary, it will be shown that symbolic models are very useful for on-line applications. Unfortunately, the generation of symbolic equations is an extremely complex problem. That is the main reason for which these methods have not been developed over the past years.

At the end of this chapter, we will explain the essential concepts of a new class of methods which we have conditionally named numeric-symbolic methods and which represent a specific "hybridization" of numerical and symbolic modelling. It will be shown that the development of this class of methods yields extremely attractive results, especially in the domain of real-time applications of dynamic robot models.

1.2. Methods Based on Lagrange's Equations

In this section we will briefly present methods for the modelling of robot systems based on second-order Lagrange's equations. These equations are of the form

$$\frac{d}{dt}\left(\frac{\partial E_k}{\partial \dot{q}_i}\right) - \frac{\partial E_k}{\partial q_i} + \frac{\partial E_p}{\partial q_i} = P_i, \qquad i=1,\ldots,n \qquad (1.2.1)$$

where: n - is the number of degrees of freedom of the system,

     E_k - kinetic energy,

     E_p - potential energy,

     P_i - generalized force, and

     q_i - generalized joint coordinate corresponding to the i-th degree of freedom.

The kinetic and potential energy is obtained as a sum of the corresponding mechanism links energy. In all methods which will be presented a robot mechanism means an open kinematic chain consisting of n rigid bodies interconnected by n joints. The generalized coordinate q_i represents the internal coordinate of the i-th joint.

Although all methods to be described in this section have resulted from Lagrange's equations (1.2.1), we will see in the text to follow that each of them has a number of specific features.

1.2.1. Uicker-Kahn's method

Uicker-Kahn's method is one among the first methods for constructing dynamic robot models. In its original form it was developed in 1965 by J.J. Uicker [1-3]. Since this method related mainly to certain classes of closed mechanisms, in 1969 M. Kahn [4, 5] elaborated an algorithm for modelling open kinematic chains. In 1971 the method was modified and preformulated by introducing "screw" vectors [6, 7] by L. Woo and F. Freundenstein [8] and A. Yang [9]. In 1981 N. Orleandea and T. Berenyi [10] gave the program implementation of this method in the form of a program package for dynamic analysis of robots. Improved variants of this method in the sense of reducing the number of numerical operations were elaborated by S. Mahil [11, 12], M. Renaud [13], M. Tomas and D. Tesar [14], R. Waters [15], J. Hollerbach [16, 17], etc.

Using the Uicker-Kahn's method we can solve either the direct or the inverse problem of dynamics. Moreover, the method enables the calculation of all matrices of dynamic robot model: the inertial matrix, the matrix of Coriolis and centrifugal effects and the gravity vector. The dynamic equations of an n-degree-of-freedom manipulator, derived using this method, have the following form [1-5]:

$$
P_i = \sum_{j=i}^{n} \left\{ \sum_{k=1}^{j} \left[tr\left(\frac{\partial W_j}{\partial q_i} J_j \frac{\partial W_j^T}{\partial q_k}\right) \right] \ddot{q}_k \right.
$$

$$
\left. + \sum_{k=1}^{j} \sum_{\ell=1}^{j} \left[tr\left(\frac{\partial W_j}{\partial q_i} J_j \frac{\partial^2 W_j^T}{\partial q_k \partial q_\ell}\right) \dot{q}_k \dot{q}_\ell \right] - m_j \vec{g}^T \frac{\partial W_j}{\partial q_i} \tilde{r}_{jo} \right\} \qquad (1.2.2)
$$

where: P_i - is driving force acting at the i-th joint if the joint is
             a sliding one, or driving torque if the i-th joint is a
             revolute one,

W_i — transformation matrix between the i-th local coordinate system and the reference system,

J_i — inertia matrix of the i-th link with respect to local coordinate system,

m_i — mass of the link i,

$\tilde{r}_{io}$ — the distance vector between the center of mass of the link i and the origin of reference coordinate system, expressed in the local coordinate system of the i-th link, and

$\vec{g}$ — gravity vector.

The matrix W_i may be expressed as

$$W_i = A_1^o A_2^1 \cdots A_i^{i-1} \tag{1.2.3}$$

where A_k^{k-1} is a (4×4) — transformation matrix between coordinate system $O_k x_k y_k z_k$ attached to link k into system $O_{k-1} x_{k-1} y_{k-1} z_{k-1}$ attached to link k-1. Let $v^k = \begin{bmatrix} 1 & x^k & y^k & z^k \end{bmatrix}$ be an arbitrary vector expressed with respect to the coordinate system of link k. For adjacent coordinate systems it holds

$$v^{k-1} = A_k^{k-1} v^k$$

From the definition of matrix W_i (1.2.3) we see that this matrix really represents the matrix of transformation of coordinates from the i-th into the reference coordinate system Oxyz. We see that Uicker has accepted 4×4 transformation matrices. Thus, both rotation and translation between links can be described in an unique way. This provides for a consistent treatment of linkage mechanisms which include revolute and sliding joints. Uicker has accepted these matrices in accordance with Denavit-Hartenberg's notation [18] which is commonly used in the literature (see, for example, [12, 13, 17]). According to this notation, the matrix of transformation A_k^{k-1} has the form:

$$A_k^{k-1} = \begin{bmatrix} 1 & 0 & 0 & 0 \\ a_k \cos q_k & \cos q_k & -\sin q_k \cos \alpha_k & \sin q_k \sin \alpha_k \\ a_k \sin q_k & \sin q_k & \cos q_k \cos \alpha_k & -\cos q_k \sin \alpha_k \\ s_k & 0 & \sin \alpha_k & \cos \alpha_k \end{bmatrix} \tag{1.2.4}$$

where: a_k - is distance between the origins of local coordinate systems
of the link $(k-1)$ and link k, measured along x_k axes (Fig.
1.1),

$\quad$ s_k - distance between axes x_{k-1} and x_k measured along z_{k-1},

$\quad$ α_k - angle between axes z_{k-1} and z_k measured along axis x_k in
the right sense, and

$\quad$ q_k - angle between axes x_{k-1} and x_k measured about axis z_{k-1} in
the right sense. The angle q_k is also called "joint coordi-
nate" of the k-th joint, if it is a revolute one. In case
of a sliding joint, the length s_k represents the joint coordinate.

The axes of the local coordinate system of the link k are defined by
unit vectors $\vec{x}_k$, $\vec{y}_k$ and $\vec{z}_k$, where:

$\quad$ $\vec{z}_k$ - is in the direction of the joint axis between links k and
$(k+1)$. By definition, this joint is regarded as the joint
$(k+1)$,

$\quad$ $\vec{x}_k$ - the vector along the line perpendicular to axes z_{k-1} and z_k,
directed to z_k, and

$\quad$ $\vec{y}_k$ - the unit vector which forms a right-handed set of vectors
$\left(\vec{x}_k, \vec{y}_k, \vec{z}_k\right)$.

It may be seen that the choice of coordinate systems depends only on
the geometrical mechanism parameters. Consequently, the axes of local
systems do not coincide with the principal inertia axes. Thus, instead
of 3 moments of inertia, one must use the inertia tensor (9 terms).
Uicker has accepted J_i to be of 4×4 dimension, so that is includes both
the inertia tensor and the link mass [1-5], [12-17]:

$$J_i = \begin{bmatrix} m_i & m_i x_{io} & m_i y_{io} & m_i z_{io} \\ m_i x_{io} & \frac{1}{2}\left(-I_{xx}+I_{yy}+I_{zz}\right) & I_{xy} & I_{xz} \\ m_i y_{io} & I_{xy} & \frac{1}{2}\left(I_{xx}-I_{yy}+I_{zz}\right) & I_{yz} \\ m_i z_{io} & I_{xz} & I_{yz} & \frac{1}{2}\left(I_{xx}+I_{yy}-I_{zz}\right) \end{bmatrix}$$

where: x_{io}, y_{io} and z_{io} - are coordinates of mass center of link i with
respect to the reference system,

$$I_{xx},\ I_{yy},\ldots$$ - moments of inertia about the x_i, y_i and z_i axes of the link.

The inertial matrix J_i describes the mass distribution of link i, but also depends on the position of its center of mass with respect to the fixed system.

From the transformation matrix A_k^{k-1} we see that its partial derivative with respect to q_k may be analytically determined. The same holds for matrix W_i. Since recursive relations are not employed, this method belongs to the class of nonrecursive methods. Let us now explain in more datails how the direct and inverse problems of dynamics are solved. To

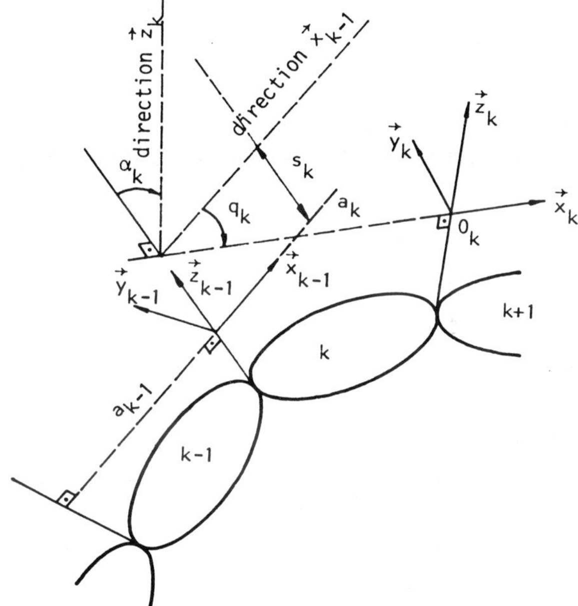

Fig. 1.1. Denavit-Hartenberg's notation

this end, let us note that joint accelerations $\ddot{q}_k$ and velocities $\dot{q}_k$ figure explicitly in model (1.2.2). Thus, it may be written in the form

$$P_i = \sum_{j=1}^{n} H_{ij}(q)\ddot{q}_j + \sum_{k=1}^{n}\sum_{\ell=1}^{n} C_{k\ell}^i(q)\dot{q}_k\dot{q}_\ell + g_i(q) \qquad (1.2.5)$$

where H_{ij}, $C_{k\ell}^i$ and g_i are scalar functions of the vector of joint coordinates $q=\begin{bmatrix} q_1 & \cdots & q_n \end{bmatrix}^T$. To prove this, let us note that, according to (1.2.2), these functions depend on the partial derivatives of matrices W_i. Since matrices W_i are functions of the joint coordinates only, it follows that the partial derivatives of W_i depend also on joint coordinates only. The model (1.2.5) may now be represented in the matrix form

$$P = H(q)\ddot{q} + \dot{q}^T C(q)\dot{q} + g(q) \qquad (1.2.6)$$

where: $P = \begin{bmatrix} P_1 & \cdots & P_n \end{bmatrix}^T$ - is the vector of driving forces (torques),

$\qquad H(q)$ - an $n \times n$ matrix called the inertial matrix of the system,

$\qquad C(q)$ - an $n \times n \times n$ matrix called the matrix of Coriolis and centrifugal effects or, shorter "C - matrix", and

$\qquad g(q)$ - an n-vector due to the action of gravity forces, which is therefore called the vector of gravity effects.

In the model (1.2.6) $\dot{q}^T C(q)\dot{q}$ denotes the n-vector

$$\dot{q}^T C(q)\dot{q} = \begin{bmatrix} \dot{q}^T C^1(q)\dot{q} \\ \vdots \\ \dot{q}^T C^n(q)\dot{q} \end{bmatrix}$$

where $C^i(q)$ are $n \times n$ matrices whose elements correspond to $C^i_{k\ell}(q)$ in (1.2.5). Frequently, the model (1.2.6) is also written in the form

$$P = H(q)\ddot{q} + h(q, \dot{q}) \qquad (1.2.7)$$

where $h(q, \dot{q}) = \dot{q}^T C(q)\dot{q} + g(q)$. We can now directly solve the inverse problem of dynamics, i.e. to determine driving forces and moments (P) based on the known motion $(q, \dot{q}, \ddot{q})$. To solve the direct problem of dynamics, it is necessary to determine $q(t)$, $t > t_o$, for known $P(t)$, $t > t_o$, and initial conditions $q(t_o)$ and $\dot{q}(t_o)$. Obviously, the model (1.2.7) yields the solution to the problem since the vector $\ddot{q}$ figures explicitly, and the inertial matrix $H(q)$ has the inverse matrix[*]. Thus, we obtain

$$\ddot{q} = H(q)^{-1}(P - h(q, \dot{q})) \qquad (1.2.8)$$

---

[*] This property will be proved in Chapter 2.

wherefrom $q(t)$, $t > t_o$, is obtained using any integration method.

Finally, let us consider the number of numerical operations required to form the dynamic robot model by Uicker-Kahn's method. Let us note immediately that by numerical operations we mean the multiplication and addition (subtraction) of real numbers (floating-point numbers). In this method, as well as in the remaining ones which will be presented later, other operations also appear, e.g., cosine and sine functions. However, these functions must be calculated in any method. This is why our comparison of different methods will be based on the number of multiplications and additions, with the addition meaning either addition or subtraction of real numbers. The remaining operations figuring in program implementation (such as transfer of contents from one memory location to another, logic functions, program calls, etc.) will not be taken into account, since this would require a very complex analysis of methods, and, on the other hand, these operations mainly depend on the properties of the programming language used.

In paper [17] it is shown that the number of multiplications and additions required to form the dynamic model by Uicker-Kahn's method depends on n^4; thus, for 3-degree-of-freedom manipulators, it amounts to over five thousand multiplications and approximately as many additions. For 6-degree-of-freedom manipulators, the number of operations increases by more than ten times. The question arising is concerned with the possibilities of real-time implementation on today's microcomputers. In the literature, a model is usually regarded as real-time implementable if it can be calculated at least each 20 ms, which corresponds to 50 Hz. As has been stated in the introduction, the question whether an algorithm is real-time implementable or not is closely related to a referent computer on which the algorithm is processed. Further, it will be accepted that contemporary 16-bit microcomputers (with arithmetic coprocessors or appropriate hardware support) execute one multiplication and one addition in about 0.1 ms. It is now possible to estimate the time required to form the model (1.2.2). More than 500 ms is obtained for 3-degree-of-freedom, and over 5 s for 6-degree-of-freedom manipulators. As may be seen, this time is more than 25 times larger than admissible (20 ms) for 3-degrees-of-freedom, and 250 times larger for 6-degrees-of-freedom. This has given rise to a series of preformulations of the basic method intended to reduce the number of operations. The basic ideas and numerical efficiency of these algorithms will be considered in this section, without giving any detailed descriptions.

1.2.2. <u>Algorithms by S. Mahil, S. Megahed</u>
<u>and M. Renaud</u>

From the foregoing text, we see that - the second-order partial derivatives figure in Uicker-Kahn's method. This is the main disadvantage of this method in the sense of efficient program implementation in real time. This is why S. Mahil [11, 12], and S. Megahed and M. Renaud [19] gave a reformulation of Uicker-Kahn's method intended to eliminate the partial derivatives and reduce the number of numerical operations.

In 1979 S. Mahil proposed Rodrigues' finite-rotations formula to be used instead of Denavit-Hartenberg's matrices. This formula was first employed in Newton-Euler's method in 1973 [20-22]. Correlation between Denavit-Hartenberg's matrices and Rodrigues' formula is described in detail in paper [23]. Although the Rodrigues' formula will be presented in detail in Ch. 2, we will briefly describe it here because of its importance. Let us consider the rotation of the i-th link with respect to the (i-1)-st by angle q_i (Fig. 1.2). Let us denote by $\vec{r}_i$ an arbitrary vector attached to the i-th link before the rotation by q_i. After the rotation according to the Rodrigues' formula this vector may be expressed as

$$\vec{r}'_i = \vec{r}_i \cos q_i + \left(1-\cos q_i\right)\left(\vec{e}_i \cdot \vec{r}_i\right)\vec{e}_i + \vec{e}_i \times \vec{r}_i \sin q_i \qquad (1.2.9)$$

where $\vec{e}_i$ is the unit vector of the i-th joint axis. Let us now apply this formula to determining the transformation matrix A_i^{i-1} which describes rotation of the i-th coordinate system with respect to the (i-1)-st. Let us denote by $\vec{q}_{ij}$, (j=1,2,3), the unit vectors of the i-th coordinate system axes with respect to the (i-1)-st system. The matrix A_i^{i-1} is then given by

$$A_i^{i-1} = \left[\vec{q}_{i1} \quad \vec{q}_{i2} \quad \vec{q}_{i3}\right] \qquad (1.2.10)$$

To determine the vector $\vec{q}_{ij}$, we will apply the Rodrigues' formula

$$\vec{q}_{ij} = \vec{q}^{\,o}_{ij} \cos q_i + \left(1-\cos q_i\right)\left(\vec{e}_i \cdot \vec{q}^{\,o}_{ij}\right)\vec{e}_i + \vec{e}_i \times \vec{q}^{\,o}_{ij} \sin q_i \qquad (1.2.11)$$

where $\vec{q}^{\,o}_{ij}$ is the vector of the axis j of the i-th coordinate system before rotation. Since vectors $\vec{e}_i$ and $\vec{q}^{\,o}_{ij}$ with respect to the (i-1)-th system are known, formula (1.2.7) directly determines the transformation matrix A_i^{i-1}.

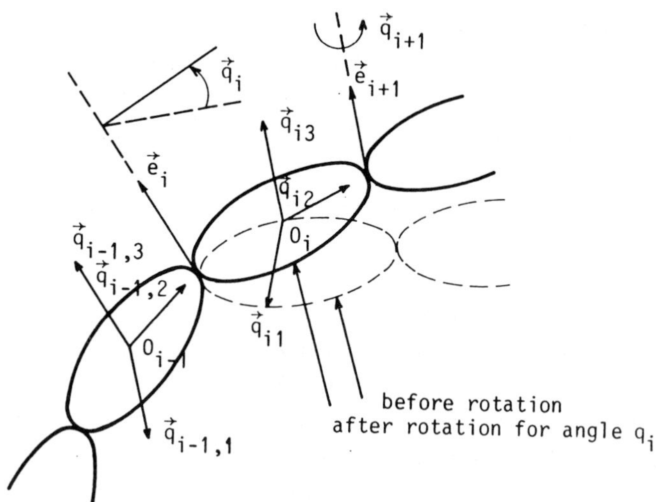

Fig. 1.2. Rotation in joint i

In contrast to Denavit-Hartenberg's kinematics, the Rodrigues' formula applies to local coordinate systems which are attached to links in an arbitrary way. It is most convenient to accept that the coordinate origin coincides with the mass center of the link and that the axes are directed along the principal axes of inertia. Dynamic equations are thus considerably simplified, since the principal moments of inertia are used instead of inertia tensor.

Let us now return to the algorithm by S. Mahil. In this algorithm no more than 2 moments of inertia figure - longitudinal and transversal. However, this approximation is justifiable only for cylindrical links whose lengths are considerably larger than diameters. Unfortunately, such an approximation is often unjustifiable.

Starting from Lagrange's equations, S. Mahil obtains dynamic equations in the matrix form

$$P = \sum_{i=1}^{n} \left(H^i(q)\ddot{q} + D^i(q, \dot{q})\dot{q} + g^i(q) \right) \qquad (1.2.12)$$

where: $P \in R^{n}$ *) - is the vector of driving forces (moments),

$\quad\quad$ $H^i(q)$ and $D^i(q, \dot{q})$ - n×n matrices, and

$\quad\quad$ $g^i(q)$ - an n-vector

The elements of the matrices in (1.2.12) may be expressed using the vectors of joint axes, local coordinate systems, and distance vectors. These vectors are easily obtained applying Rodrigues' formula. To illustrate this method, we will present a number of characteristic expressions for the elements of model matrices. For example, in the case of revolute joints, the (j, k)-th element of matrix $H^i(q)$ may be expressed as

$$H^i_{jk} = \left(Jn_i + m_i\left(\vec{r}_{ji} \cdot \vec{r}_{ki}\right)\right)\left(\vec{e}_j \cdot \vec{e}_k\right) + Js_i\left(\vec{e}_j \cdot \vec{q}_{i1}\right)\left(\vec{e}_k \cdot \vec{q}_{i1}\right) -$$

$$- m_i\left(\vec{e}_j \cdot \vec{r}_{ki}\right)\left(\vec{e}_k \cdot \vec{r}_{ji}\right), \quad\quad (j<k<i), \quad\quad\quad (1.2.13)$$

where: Jn_i - the transversal moment of inertia of the i-th link,

$\quad\quad$ Js_i - the longitudinal moment of inertia of the i-th link,

$\quad\quad$ m_i - mass of the i-th link,

$\quad\quad$ $\vec{r}_{ij}$ - the vector of distance from the j-th joint to the mass center of the i-th link, expressed in the inertial system.

Vectors $\vec{e}_i$ and $\vec{q}_{ij}$ are also expressed in the inertial system.

The elements of matrices $D^i(q, \dot{q})$ may be expressed in the form:

$$D^i_{k\ell} = \sum_{j=p+1}^{i} \frac{\partial H^i_{k\ell}}{\partial q_j} \dot{q}_j - \frac{1}{2} \sum_{j=1}^{q} \frac{\partial H^i_{j\ell}}{\partial q_k} \dot{q}_j \quad\quad\quad (1.2.14)$$

where p and q depend on k and ℓ [12]. This expression includes the partial derivatives of the terms of matrices which should be formed in closed form **). Since these expressions are rather complex and depend on the relation between indices i, k, ℓ, and j, for illustration we will give here only one of the expressions applying to revolute joints under the condition $k<j<\ell<i$:

---

*) $(\cdot) \in R^{n}$ means that $(\cdot)$ is an n-vector whose elements are real numbers.

**) Here, the term closed form means an expression whose value is not obtained from one or more recursive relations, but by directly substituting its arguments by appropriate numerical values.

$$\frac{\partial H_{k\ell}^{i}}{\partial q_{j}} = \left(Jn_{i} + m_{i}\vec{r}_{ki} \cdot \vec{r}_{\ell i} \right)\left(\vec{e}_{k} \cdot \vec{e}_{j} \times \vec{e}_{\ell} \right) +$$
$$+ m_{i}\left[\vec{e}_{k} \cdot \vec{e}_{\ell}\left(\vec{R}_{kj} \cdot \vec{e}_{j} \times \vec{r}_{\ell i} \right) - \vec{e}_{\ell} \cdot \vec{r}_{ki}\left(\vec{R}_{kj} \cdot \vec{e}_{j} \times \vec{e}_{\ell} \right) -$$
$$- \vec{e}_{k} \cdot \vec{r}_{\ell i}\left(\vec{R}_{kj} \cdot \vec{e}_{j} \times \vec{e}_{\ell} \right) \right] + Js_{i}\vec{e}_{\ell} \cdot \vec{q}_{ik}\left(\vec{e}_{k} \cdot \vec{e}_{j} \times \vec{q}_{i1} \right) \qquad (1.2.15)$$

where $\vec{R}_{kj}$ is the vector of distance from the mass center of the link j to
the mass center of the link k.

Expression (1.2.14) relates the elements of the matrices of Coriolis
and centrifugal effects to the partial derivatives of the elements of
inertial matrix. However, this expression is complicated and "clumsy"
to a certain extent. A more elegant form of this expression will be derived in Ch. 2. Expressions (1.2.13) and (1.2.15) will be derived for
links of arbitrary form. This is why these expressions will not be considered in detail here.

It follows from the preceding text that this method belongs to the class
of nonrecursive methods based on Lagrange's equations. By comparing it
to Uicker-Kahn's method, one may see that the number of numerical operations in these two methods is of the same order of magnitude. Thus, it
follows that, in spite of the approximations introduced in this method,
it is practically impossible to implement it in real time. Nevertheless,
the main quality of the algorithm developed lies in that it provides a
direct insight into the physical parameters figuring in the elements of
dynamic model matrices.

In 1982 M. Thomas and D. Tesar [14] presented an algorithm which is, in
its basic idea, similar to the previously described one. This algorithm
is also intended to determine closed-form expressions for the elements
of dynamic model matrices, starting from Lagrange's equations. The main
difference lies in the fact that this algorithm employs "rotation matrices" instead of Rodrigues' formula. The rotation matrices are defined with respect to local coordinate systems which correspond to Denavit-Hartenberg's notation (Fig. 1.1). They represent transformation matrices from
local coordinate systems into systems with the same coordinate origins,
but with axes parallel with the axes of the reference coordinate frame.
In contrast to Denavit-Hartenberg's 4×4 matrices, rotation matrices are
of 3×3 dimensions. M. Thomas and D. Tesar [14] have shown that the matrix
of rotation of the k-th link may be expressed as

20

$$A_k = \left[\vec{a}_k^{\,o} \;\middle|\; \vec{z}_{k-1} \times \vec{a}_k^{\,o} \;\middle|\; \vec{z}_{k-1} \right] \qquad (1.2.16)$$

where $\vec{a}_k^{\,o}$ is the unit vector of the line perpendicular to axes $\vec{z}_{k-1}$ and $\vec{z}_k$, directed to $\vec{z}_k$ (see Fig. 1.1).

Starting from kinematics defined in this way, the authors have derived a series of matrix expressions for dynamics robot analysis. However, the authors have not analyzed the advantages of this algorithm with respect to Uicker-Kahn's method.

In 1981 and 1982 S. Megahed and M. Renaud [19] proposed an efficient procedure for calculating the matrices of partial derivatives which figure in Uicker-Kahn's method. The algorithm is based on the following property of Denavit-Hartenberg's matrices

$$\frac{\partial A_k^{k-1}}{\partial q_k} = Q A_k^{k-1} \qquad (1.2.17)$$

where

$$Q = \begin{bmatrix} 0 & 0 & 0 & 0 \\ 0 & 0 & -1 & 0 \\ 0 & 1 & 0 & 0 \\ 0 & 0 & 0 & 0 \end{bmatrix}$$

if the joint k is a revolute one, and

$$Q = \begin{bmatrix} 0 & 0 & 0 & 0 \\ 0 & 0 & 0 & 0 \\ 0 & 0 & 0 & 0 \\ 1 & 0 & 0 & 0 \end{bmatrix}$$

if joint k is a sliding one. This property is easily checked by differentiation (1.2.4). On the basis of this relation, it was shown that the following holds

$$\frac{\partial W_j}{\partial q_i} = \left[\begin{array}{c|c} 0 & \\ \hline \Omega_i & \end{array} \right] W_j, \qquad (i<j)$$

$$\frac{\partial W_j}{\partial q_i \partial q_k} = \left[\begin{array}{c|c} 0 & \\ \hline \Omega_i & \end{array} \right]\left[\begin{array}{c|c} 0 & \\ \hline \Omega_k & \end{array} \right] W_j, \qquad (i<k<j)$$

$$(1.2.18)$$

where Ω_i is a 3×3 matrix whose elements are directly obtained from the matrix W_j, [19].

From the foregoing discussion, we see that this algorithm belongs to the class of nonrecursive algorithms based on Lagrange's equations. However, the number of numerical operations is still very large thus making this algorithm unsuitable for real-time application.

Finally, let us mention that M. Renaud presented in 1983 yet another modification of the previous method, by introducing tensor calculus, as well as a series of recursive relations for calculating the elements of dynamic model matrices [24]. But, neither this modification has resulted in a considerable reduction of the number of numerical operations. However, the author has shown in this paper that the analytical model obtained by applying this procedure is very efficient in the numerical sense. Unfortunately, the procedure for forming the analytical model is not automated, so it cannot be implemented on a computer but is to be performed "by hand". This procedure is tedious and complex and errors are almost inevitable. Therefore, this procedure cannot be regarded as computer-oriented, and cannot be compared with the other computer-oriented algorithms.

1.2.3. Algorithms by R. Waters and J. Hollerbach

In 1979 and 1980 R. Waters [15] and J. Hollerbach [16, 17] developed algorithms for solving the inverse problem of dynamics, as a specific case of Uicker-Kahn's method. The procedures developed are used to determine the driving forces (torques) on the basis of known motion. These procedures do not enable computation of inertial system matrix, since the second derivatives of joint coordinates implicitly figure in these algorithms. This is why these algorithms are not equivalent to Uicker-Kahn's method, and the performances of these methods are comparable only when the inverse problem of dynamics is concerned.

According to the algorithm by R. Waters, the driving forces and torques can be expressed as

$$P_i = \sum_{j=i}^{n} \left[\mathrm{tr}\!\left(\frac{\partial W_j}{\partial q_i} J_j \ddot{W}_j^T\right) - m_j \vec{g}^T \frac{\partial W_j}{\partial q_i} \tilde{r}_{jo} \right] \qquad (1.2.19)$$

with recursive relations for W_j, $\dot{W}_j$ and $\ddot{W}_j$ being

$$W_j = W_{j-1} A_j^{j-1}$$

$$\dot{W}_j = \dot{W}_{j-1} A_j^{j-1} + W_{j-1} \frac{\partial A_j^{j-1}}{\partial q_j} \dot{q}_j \tag{1.2.20}$$

$$\ddot{W}_j = \ddot{W}_{j-1} A_j^{j-1} + 2\dot{W}_{j-1} \frac{\partial A_j^{j-1}}{\partial q_j} \dot{q}_j + W_{j-1} \frac{\partial^2 A_j^{j-1}}{\partial q_j^2} \dot{q}_j^2 + W_{j-1} \frac{\partial A_j^{j-1}}{\partial q_j} \ddot{q}_j$$

These recursive relations reduce the number of multiplications/additions required to compute the driving forces to the dependence on n^2 (for example, the number of additions equals $82n^2 + 514n - 384$). The reduced number of operations as compared with Uicker-Kahn's method evidently results from the fact that the inertial system matrix is not explicitly calculated, so it is not required to compute the partial derivatives of $\partial W_j / \partial q_k \partial q_\ell$ type. However, in spite of these numerical savings, the number of operations is still very large for real-time application (for example, for $n = 6$, the number of multiplications amounts to over 7000, [16]).

A significant reduction of the number of operations was obtained by J. Hollerbach [16, 17], who recognized that the partial derivative $\partial W_j / \partial q_i$ may be expressed as $\left(\partial W_i / \partial q_i \right) W_j^i$, where

$$W_j^i = A_{i+1}^i A_{i+2}^{i+1} \cdots A_j^{j-1} \tag{1.2.21}$$

is the matrix of transformation from the i-th into the j-th local coordinate system. Substituting the foregoing expression into (1.2.19), the dynamic model is obtained in the form

$$P_i = \text{tr}\left(\frac{\partial W_i}{\partial q_i} D_i \right) - \vec{g}^T \frac{\partial W_i}{\partial q_i} c_i \tag{1.2.22}$$

with $D_i \in R^{4 \times 4}$ and $c_i \in R^4$ calculated from the following recursive relations

$$D_i = J_i \ddot{W}_i^T + A_{i+1}^i D_{i+1},$$

$$c_i = m_i \tilde{r}_{io} + A_{i+1}^i c_{i+1}. \tag{1.2.23}$$

Accelerations $\ddot{W}_i^T$ are calculated by forward recursion (from $i = 1$ to n), as in the Waters' algorithm. D_i and c_i are calculated by backward recursion (from $i = n$ to $i = 1$). Thus, one obtains that the number of multiplications is $n_M = 830n - 592$, and the number of additions $n_A = 675n - 464$.

Hollerbach has also recognized that it is possible to reduce the number of operations further, by using 3×3 - instead of 4×4 - matrices. As we have seen, 4×4 - matrices simultaneously describe rotation and translation thus introducing considerable computational redundancy, since rotation may be described by a 3×3 - matrix and translation by a position vector. Since this preformulation of the algorithm results in no essentially new results compared with the previously described ones, we will present only the number of operations required to form the driving forces (torques):

$$n_M = 412n - 277, \qquad n_A = 320n - 201.$$

It will be shown, however, that Newton-Euler's approach may give an even smaller number of operations.

1.2.4. <u>Vukobratović-Potkonjak's recursive method</u>

The methods based on Lagrange's equations enabling either the direct or the inverse problem of dynamics to be solved have been presented in Subsections 1.2.1 and 1.2.2. Closed-form expressions are employed in these methods, and no recursive relations are used. However, the question arises whether it is possible to obtain the dynamic model matrices by recursive expressions, in order to reduce the number of operations. This problem was solved in 1979 by Vukobratović and Potkonjak, as described in [25]. This method is also presented in detail in Ref. [26].

The method employs Rodrigues' formula to describe the spatial disposition of links. Since this approach has already been described in 1.2.2, we will only emphasize the fact that such an approach facilitates the dynamic analysis of a mechanism, since the principal moments of inertia are used instead of inertia tensor.

The kinematic part of this method reduces to the following two basic relations

$$\tilde{\omega}_i = \tilde{N}(i)\dot{q}$$

$$\tilde{v}_i = \tilde{M}(i)\dot{q} \tag{1.2.24}$$

where: $\tilde{\omega}_i$ - vector of the i-th link angular velocity expressed with re-
    spect to the i-th local coordinate system,

  $\tilde{v}_i$ - vector of the linear velocity of the i-th link mass center
    with respect to the i-th local coordinate system,

 $\tilde{N}(i)$, $\tilde{M}(i)$ - 3×n matrices depending on joint coordinates, formed by
    recursive relations [25],

 $\dot{q}$ - n×1 vector of joint velocities

Substituting (1.2.24) into the expression for the kinetic energy of the overall system

$$E_k = \sum_{i=1}^{n} \left(\frac{1}{2} m_i \tilde{v}_i^T \tilde{v}_i + \frac{1}{2} \tilde{\omega}_i \tilde{J}_i \tilde{\omega}_i \right) \tag{1.2.25}$$

and using the matrix form of Lagrange's equations (1.2.1)

$$\frac{d}{dt}\left(\frac{\partial E_k}{\partial \dot{q}}\right) - \frac{\partial E_k}{\partial q} + \frac{\partial E_p}{\partial q} = P \tag{1.2.26}$$

the dynamic robot model is obtained in the form (1.2.7)

$$P = H(q)\ddot{q} + h(q, \dot{q}).$$

In the above expressions m_i denotes the i-th link mass, $\tilde{J}_i$ - 3×3 diagonal matrix whose elements are the principal moments of inertia, E_k and
E_p - the kinetic and potential energy of the overall system, and P -
the vector of driving forces (torques). The model matrices are given by
the expressions

$$H(q) = \sum_{i=1}^{n} \left(m_i \tilde{M}(i)^T \tilde{M}(i) + \tilde{N}(i)^T \tilde{J}_i \tilde{N}(i) \right) \tag{1.2.27}$$

$$h(q, \dot{q}) = \frac{\partial \left(E_p - E_k \right)}{\partial q} + \dot{H}(q)\dot{q} \tag{1.2.28}$$

The partial derivatives in the latter expression, as well as the time
derivative of the inertial matrix, $H(q)$, are also calculated by recursive relations [25], which will not be presented here because of their
complexity.

The number of numerical operations obtained by applying this method is
several times less than that in Uicker-Kahn's method. However, even
this number is still large comparing to Newton-Euler's method.

1.3. Methods Based on Newton-Euler's Equations

In this section we will present some robot modelling algorithms based on Newton-Euler's equations. Together with Lagrange's these algorithms are most commonly used in robot modelling.

Newton-Euler's equations determine the inertial forces and torques acting on links. If we denote ty $\tilde{F}_i$ the inertial force acting to the i-th link mass center, expressed in the local coordinate system of that link, and by m_i the i-th link mass, Newton's law yields

$$\tilde{F}_i = m_i \tilde{w}_i \qquad (1.3.1)$$

where $\tilde{w}_i$ is the acceleration of the i-th link mass center with respect to i-th link coordinate system. Euler's dynamic equation determines the moment of inertial forces about the i-th link mass center

$$\tilde{M}_i = \tilde{J}_i \tilde{\varepsilon}_i + \tilde{\omega}_i \times \left(\tilde{J}_i \tilde{\omega}_i \right) \qquad (1.3.2)$$

where: $\tilde{J}_i$ - 3×3 matrix whose diagonal elements are the principal moments of inertia,

      $\tilde{\varepsilon}_i$ - angular acceleration of the i-th link, and

      $\tilde{\omega}_i$ - angular velocity of the i-th link.

All quantities in (1.3.2) are expressed in the local coordinate system of the i-th link, placed at the mass center. Local system axes coincide with the central axes of inertia.

When expressed with respect to the fixed, inertial system, Newton-Euler's equations have the same form as (1.3.1) and (1.3.2), with $J_i = A_i \tilde{J}_i A_i^T$ figuring instead of $\tilde{J}_i$. A_i is the matrix of transformation from the local coordinate system of the i-th link into the fixed system. J_i is also a 3×3 matrix and represents the inertia matrix of the i-th link with respect to the reference coordinate frame.

In the following text we will present Newton-Euler's methods developed with respect to the fixed, inertial coordinate system, as well as the methods derived with respect to the local coordinate systems.

1.3.1. Vukobratović-Stepanenko's method

Newton-Euler's dynamic equations were first applied to modelling of active mechanisms in 1973 by M. Vukobratović and J. Stepanenko [20]. The concept of recursiveness in solving either the direct or the inverse problem of dynamics was also introduced in this paper. This method has also been referred to as kinetostatic method, method based on D'Alembert's principle, or method of general theorems of mechanics [21, 22]. A slightly more efficient, computer oriented procedure based on this method was elaborated in [27 - 29]. Introducing Euler's parameters, R. Huston and F. Kelly [30] derived a nonrecursive procedure for dynamic modelling, which is essentially based on Newton-Euler's dynamic equations and T. Kane's algorithm [31, 32]. Another variant of Newton--Euler's method intended for efficient solution to the inverse problem (i.e., determination of driving forces, given a manipulator motion) was elaborated by J. Luh, M. Walker and R. Paul [33]. Newton-Euler's equations were also used by A. Leskov and B. Medvedov [34] in the method of "block-matrices". This method is presented in the Monograph by E. P. Popov and associates [35]. Since the same method is described in detail in Volume 1 of this series, it will not be considered here. Newton-Euler's method was the basis for elaborating the algorithm for computing linearized dynamic equations of open kinematic chains [36 - 38] as well as the algorithm for sensitivity equations computation [39 - 40].

Just as the methods based on Lagrange's equations, Newton-Euler's methods employ either Rodrigues' formula [20 - 22] or Denavit-Hartenberg's matrices [33]. The correspondence between the two above mentioned approaches and their convenience for robot kinematics and dynamics were first presented in article [23]. Namely, in Denavit-Hartenberg's kinematic notation axes of the link coordinate frames do not coincide, in a general case, with principal inertia axes of the links, so that operation with inertia tensor is needed. On the contrary, in Rodrigues' formula, the local coordinate systems' axes coincide with the directions of the principal inertia axes, so that only 3 principal moments of inertia figure in dynamic equations, instead of the inertia tensor. In the following text we will briefly present the basic relations of Newton-Euler's method which starts from the Rodrigues' formula (this method will be described in detail in Chapter 2).

Assume that, by applying the Rodrigues' formula, the vectors which describe the spatial disposition of the links, given a set of joint coordinates $q_1,\ldots,q_n$, have been determined

$\vec{q}_{ij}$, (j=1,2,3) - unit vector of the j-th axis of the local coordinate
system of the i-th link,

$\vec{r}_{ii}$ - distance vector from the i-th joint to the i-th link
mass center

$\vec{r}_{i,i+1}$ - distance vector from the (i+1)-th joint to the i-th
link mass center,

$\vec{e}_i$ - unit vector of the i-th joint axis,

for each i=1,...,n (Fig. 1.3).

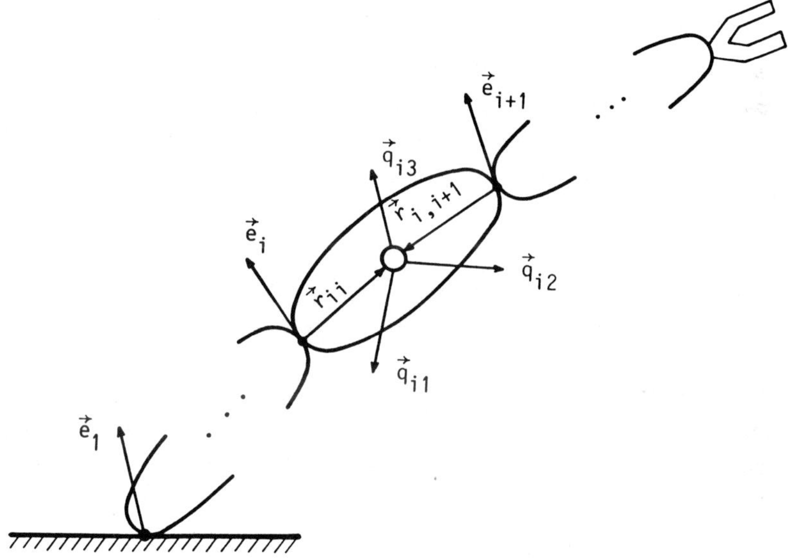

Fig. 1.3. Manipulator link

Angular velocity $\vec{\omega}_i$, angular and linear accelerations ($\vec{\varepsilon}_i$ and $\vec{w}_i$) are
expressed by the following equations

$$\vec{\omega}_i = \vec{\omega}_{i-1} + \dot{q}_i \vec{e}_i$$

$$\vec{\varepsilon}_i = \sum_{j=1}^{i} \vec{\alpha}_{ij} \ddot{q}_j + \vec{\alpha}_i^o \qquad (1.3.3)$$

$$\vec{w}_i = \sum_{j=1}^{i} \vec{\beta}_{ij} \ddot{q}_j + \vec{\beta}_i^o$$

28

where coefficients $\vec{\alpha}_{ij}$, $\vec{\alpha}_i^O$, $\vec{\beta}_{ij}$ and $\vec{\beta}_i^O$ are obtained from the corresponding recursive relations, [20 - 22]. For example, $\vec{\alpha}_i^O$ is obtained from

$$\vec{\alpha}_i^O = \vec{\alpha}_{i-1}^O + \dot{q}_i \vec{\omega}_{i-1} \times \vec{e}_i \tag{1.3.4}$$

These relations will be derived in Ch. 2. Let us note that two essential properties of this method are incorporated in these relations:

1. the property of recursiveness,

2. joint accelerations $\ddot{q}_j$ figure explicitly in these equations, thus enabling extraction of the inertial matrix and, solution of the direct dynamic problem.

The inertial force and moment may also be expressed by similar relations

$$\vec{F}_i = \sum_{j=1}^{i} \vec{a}_{ij}\ddot{q}_j + \vec{a}_i^O$$

$$\vec{M}_i = \sum_{j=1}^{i} \vec{b}_{ij}\ddot{q}_j + \vec{b}_i^O \tag{1.3.5}$$

where coefficients $\vec{a}_{ij}$, $\vec{a}_i^O$, $\vec{b}_{ij}$ and $\vec{b}_i^O$ are obtained from the corresponding recursive relations. Let us note that coefficients $\vec{b}_{ij}$ and $\vec{b}_i^O$ are obtained from Euler's dynamic equations and that they emloy only the principal moments of inertia and no inertia tensor. Finally, motion equations are obtained in the matrix form

$$P = H(q)\ddot{q} + h(q, \dot{q})$$

where $H(q)$ is an n×n inertial matrix:

$$H_{ik} = -\vec{e}_i \cdot \sum_{j=\max(i,k)}^{n} \left(\vec{b}_{jk} + \vec{r}_{ji} \times \vec{a}_{jk} \right), \qquad (i,k=1,\ldots,n) \tag{1.3.6}$$

and $h(q, \dot{q})$ a vector with n elements:

$$h_i = -\vec{e}_i \cdot \sum_{j=1}^{n} \left(\vec{r}_{ji} \times \left(\vec{a}_j^O + \vec{G}_j \right) + \vec{b}_j^O \right) \tag{1.3.7}$$

where $\vec{G}_j$ is the gravity vector of the j-th link. The preceding relations are written for revolute joints, but can easily be modified for sliding ones as well [22, 41].

The number of numerical operations needed for computing the matrices $H(q)$ and $h(q, \dot{q})$ is

$$n_M = \frac{3}{2} n^3 + 28n^2 + \frac{525}{2} n$$

$$n_A = \frac{4}{3} n^3 + 20n^2 + \frac{530}{3} n$$

where n_M denotes the number of multiplications, and n_A the number of additions.

When only the inverse problem of dynamics is solved, Newton-Euler's method becomes considerably simpler [22, 41]. Kinematics then reduces to 3 recursive relations

$$\vec{\omega}_i = \vec{\omega}_{i-1} + \dot{q}_i \vec{e}_i,$$

$$\vec{\varepsilon}_i = \vec{\varepsilon}_{i-1} + \ddot{q}_i \vec{e}_i + \dot{q}_i \left(\vec{\omega}_{i-1} \times \vec{e}_i \right), \tag{1.3.8}$$

$$\vec{w}_i = \vec{w}_{i-1} - \vec{\varepsilon}_{i-1} \times \vec{r}_{i-1,i} + \vec{\varepsilon}_i \times \vec{r}_{ii} - \vec{\omega}_{i-1} \times \left(\vec{\omega}_{i-1} \times \vec{r}_{i-1,i} \right) +$$

$$+ \vec{\omega}_i \times \left(\vec{\omega}_i \times \vec{r}_{ii} \right)$$

and dynamics to Newton-Euler's equations

$$\vec{F}_i = m_i \vec{w}_i$$

$$\vec{M}_i = J_i \vec{\varepsilon}_i + \vec{\omega}_i \times \left(J_i \vec{\omega}_i \right) \tag{1.3.9}$$

where $J_i = A_i \tilde{J}_i A_i^T$ is a (3×3) inertia matrix of the i-th link with respect to the reference system, and m_i is the i-th link mass.

Let us denote by $\vec{R}_i$ and $\vec{M}_i$ the force and moment by which the i-th link acts upon the (i-1)-st. The following recursive relations [22, 41] hold for these variables

$$\vec{R}_i = \vec{R}_{i+1} - \vec{F}_i - \vec{G}_i$$

$$\vec{M}_i = \vec{M}_{i+1} + \left(\vec{r}_{ii} - \vec{r}_{i,i+1} \right) \times \vec{R}_{i+1} + \vec{r}_{ii} \times \left(\vec{F}_i + \vec{G}_i \right) - \vec{M}_i \tag{1.3.10}$$

where $\vec{R}_{n+1}$ and $\vec{M}_{n+1}$ are the force and moment acting to the n-th link (robot gripper). These are backward recursive relations, since they are calculated for i=n, i=n-1,...,i=1. The driving torques are now calculated from

$$P_i = \vec{M}_i \cdot \vec{e}_i \qquad\qquad (1.3.11)$$

and the driving forces from $F_i = \vec{R}_i \cdot \vec{e}_i$, for a sliding joint.

The number of numerical operations needed for driving torques computation is

$$n_M = 225n, \qquad n_A = 152n.$$

J. Luh, M. Walker and R. Paul [33] reformulated the preceding algorithm by expressing all kinematic and dynamic variables with respect to link coordinate systems. This provides for a more efficient calculation of Euler's dynamic equations, reducing the number of numerical operations to [16, 17]:

$$n_M = 150n - 48, \qquad n_A = 131n - 48$$

Let us underline once again that this algorithm solves the inverse problem of dynamics.

Finally, let us investigate the possibilities for real-time implementation of Newton-Euler's algorithms on today's microcomputers. Let us first consider the algorithm for solving either the direct or the inverse dynamic problem (1.3.3 - 1.3.7). On the basis of the given expressions, one obtains that over 900 floating-point multiplications and nearly as many additions are required to be performed for 3-degree-of--freedom manipulators. For 6-degree-of-freedom manipulators, the number of multiplications amounts to over 2300. In presenting Uicker-Kahn's method, we have accepted that one multiplication and one addition take about 0.1 ms on contemporary microcomputers. This means that more than 90 ms are required to compute the model matrices for a manipulator with three degrees of freedom. However, this time is considerably longer than 20 ms, which is regarded as maximal sampling period for robotic systems. Thus, we conclude that these methods are not convenient for real time applications.

Let us now consider Newton-Euler's algorithms for solving the inverse problem of dynamics. According to the algorithm by J. Luh and others, it is obtained that the number of multiplications for 3-degree-of-freedom manipulators is $n_M \approx 400$, and $n_M \approx 850$ for 6-degree-of-freedom

robots. By analysis similar to that in the previous case, we conclude
that computation of driving forces (torques) for 3-degree-of-freedom
manipulators requires 40 ms. This time is still too long for real-time
system operation. Thus, it follows that Newton-Euler's method is also
inconvenient for real time implementation on today's microcomputers.

1.3.2. Huston-Kane's method

Huston-Kane's algorithm is based on Kane's dynamic equations. These
equations originate from the general theorems of mechanics and were
presented in several papers by Kane [31, 32, 42, 43]. Kane's algorithm
directly relies on Newton-Euler's dynamic equations. Let us state that
Kane's dynamic equations are not restricted to robotic systems but apply to a broader class of mechanisms, including, for example, space
aircrafts with several joint-connected rigid bodies [43]. R. Huston gave a concretization of Kane's results to robotic systems [43 - 45] and
elaborated the program implementation. We will briefly present Huston-
-Kane's algorithm, pointing out the similarities and differences compared with the previously described basic Vukobratović-Stepanenko's method.

This algorithm considers a robot consisting of n joint-connected rigid
bodies, each joint having 6 degrees of freedom. The relative orientation of adjacent links is described by 4 Euler's parameters

$$\varepsilon_{i\ell} = e_{i\ell}\sin\frac{q_i}{2}, \qquad \varepsilon_{i4} = e_{i4}\cos\frac{q_i}{2}, \qquad (\ell=1,2,3) \qquad (1.3.12)$$

where: $\vec{e}_i = \left(e_{i1}, e_{i2}, e_{i3}\right)$ – is the unit vector of the axis about which
the (i-1)-st local coordinate system is transformed into the i-th by a
simple rotation, q_i – rotation angle, and $e_{i4} = 1$. It should be stated
that Euler's parameters were used by E. Whittaker as early as 1937 [46].
However, it is easy to recognize the equivalence between Euler's parameters and Rodrigues' formula when joints with a single degree of freedom are concerned (and this is the most common case with robots). By
simple transformations, Rodrigues' formula (1.2.5) reduces to the form

$$\vec{r}' = \vec{r} + 2\sin^2\frac{q_i}{2}\,\vec{e}_i\times\left(\vec{e}_i\times\vec{r}\right) + 2\sin\frac{q_i}{2}\cos\frac{q_i}{2}\left(\vec{e}_i\times\vec{r}\right) \qquad (1.3.13)$$

where $\vec{r}$ is the vector before rotation, and $\vec{r}'$ the vector after rotation
by angle q_i about axis $\vec{e}_i$. Introducing $\vec{\varepsilon}_i = \vec{e}_i\sin\frac{q_i}{2}$ and $\varepsilon_{i4} = \cos\frac{q_i}{2}$
into this formula, we obtain

$$\vec{r}\,' = \vec{r} + 2\vec{\varepsilon}_i \times \left(\vec{\varepsilon}_i \times \vec{r} \right) + 2\varepsilon_{i4} \left(\vec{\varepsilon}_i \times \vec{r} \right). \qquad (1.3.14)$$

Thus, if Euler's parameters are known, it is immediately possible to write the Rodrigues' formula, and vice versa.

Take kinematic part of Huston-Kane's algorithm consists of expressing the velocities and accelerations of the links in terms of Euler's parameters. This is done using nonrecursive relations, such as

$$\vec{\omega}_i = \sum_{k=1}^{i} \hat{\omega}_i \qquad (1.3.15)$$

where $\vec{\omega}_i$ is the angular velocity of the i-th link with respect to the reference system, and $\hat{\omega}_k$ the relative angular velocity of the k-th link with respect to the (k-1)-st link. Let us note that this relation may be written in the recursive form

$$\vec{\omega}_i = \vec{\omega}_{i-1} + \hat{\omega}_i. \qquad (1.3.16)$$

Since $\vec{\omega}_i = \vec{e}_i \dot{q}_i$, we see that this relation is identical with the corresponding formula (1.3.3) in the previously described Newton-Euler's method. Thus, the only essential difference lies in the use of nonrecursive relations for computing all kinematic and dynamic variables, instead of recursive ones.

Dynamic analysis of this method is based on Newton-Euler's dynamic equations (1.3.9). Driving torques (forces) are determined in a way similar to that described in the previous section.

As may be seen from the preceding text, Huston's algorithm belongs to the class of nonrecursive methods based on Newton-Euler's equations. The advantage of this algorithm is its simple treatment of complex joints (with up to 6-degrees-of-freedom), i.e. complex manipulators having up to 6n-degrees-of-freedom (n - the number of links). However, this generality complicates the application of the algorithm to simple open mechanisms. On the other hand, robots most frequently belong to such a class of mechanisms. Since this is a nonrecursive algorithm, the number of numerical operations needed to compute the model matrices is considerably larger than in the previously described recursive method.

1.3.3. <u>New Recursive Method</u>

In [24] Renaud gave the analytical-iterative procedure for obtaining
inertial matrix in the mathematical model of manipulator dynamics, represented in the form:

$$H(q)\ddot{q} + \dot{q}^T C(q)\dot{q} + G(q) = P \qquad (1.3.17)$$

To obtain C matrix, he used the partial derivatives of the elements of
H matrix with respect to generalized coordinates, which represents a
considerable difficulty if the aim is to form a computer-aided procedure.

Using Lagrange's equations, Renaud [24] obtained the elements of H
matrix and recognized that these elements contain the tensor $\underline{K}^i$ which
represents in fact the set of dynamic parameters of the generalized
link with the property of iterativeness. The authors of this procedure
have recognized that the dynamic parameters of the generalized link
are common to the elements of H and C matrices.

Let us explain briefly the procedure. From the end to the base of manipulator, links n, n-1,...,i can be composed as generalized link [24].

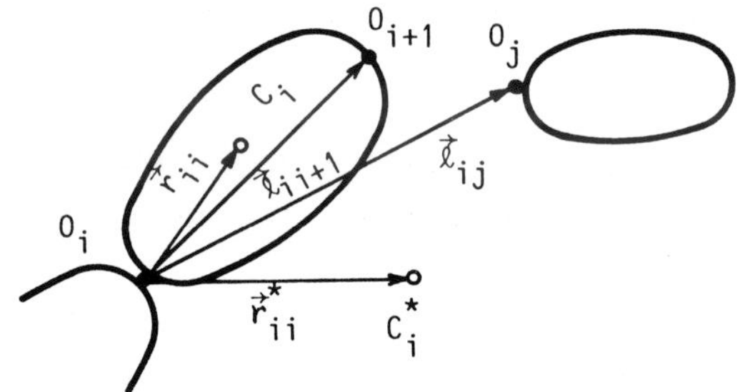

Fig. 1.4. Generalized link i

Let us introduce the symbols:

m_i^* - mass of generalized link i

$\vec{r}_{ii}^{\,*}$ - vector from origin O_i to C_i^* - mass center of generalized
      link i

$\vec{\ell}_{ij}$ - vector from origin O_i to origin O_j

$\vec{r}_{ii}$ - vector from origin O_i to C_i - mass center of link i

34

m_i - mass of link i

J_i - inertia tensor of link i with respect to origin O_i

J_i^* - inertia tensor of generalized link i with respect to origin O_i

A_i - transformation matrix between two succeding.links.

Let us introduce

$$\vec{p}_i = m_i^* \cdot \vec{r}_{ii}^* \tag{1.3.18}$$

Then we have

$$\left(J_{i+1}^*\right)_{O_i} = J_{i+1}^* + m_{i+1}^*\left(\left(\ell_{i,i+1}^2\right)I - \vec{\ell}_{i,i+1}\cdot\vec{\ell}_{i,i+1}^T\right) + \left(2\vec{p}_{i+1}^T\cdot\vec{\ell}_{i,i+1}\right)I -$$
$$- \vec{\ell}_{i,i+1}\cdot\vec{p}_{i+1}^T - \vec{p}_{i+1}\cdot\vec{\ell}_{i,i+1}^T$$

$$m_n^* = m_n, \quad m_i^* = m_{i+1}^* + m_i, \quad i=n-1,\ldots,1$$

$$\vec{p}_n = m_n\vec{r}_{nn}, \quad \left(\vec{r}_{nn}^* = \vec{r}_{nn}\right) \tag{1.3.19}$$

$$\vec{p}_i = \vec{p}_{i+1} + m_{i+1}^*\vec{\ell}_{i,i+1} + m_i\vec{r}_{ii}, \quad i=n-1,\ldots,1$$

$$J_n^* = J_n$$

$$J_i^* = J_{i+1}^* + \left(2\vec{p}_{i+1}^T\cdot\vec{\ell}_{i,i+1}\right)I - \vec{\ell}_{i,i+1}\cdot\vec{p}_{i+1}^T - \vec{p}_{i+1}\cdot\vec{\ell}_{i,i+1}^T + J_i +$$
$$+ m_{i+1}^*\left(\ell_{i,i+1}^2 I - \vec{\ell}_{i,i+1}\cdot\vec{\ell}_{i,i+1}^T\right), \quad i=n-1,\ldots,1$$

From the definitions of H, C and G matrices we have:

$$H_{ij} = \vec{e}_i^T\cdot\frac{\partial\vec{P}_i}{\partial q_j}, \quad c_{ik}^j = \frac{1}{2}\vec{e}_j^T\cdot\frac{\partial^2\vec{P}_j}{\partial q_i \partial q_k}$$

$$G_i = \vec{e}_i^T\left(\vec{r}_{ii}^* \times \left(m_i^*\vec{g}\right)\right) \tag{1.3.20}$$

The elements of these matrices in terms of the introduced dynamic parameters of generalized link are:

$$H_{ij} = \vec{e}_i^T J_i^* \vec{e}_j + \vec{e}_i^T\left(\vec{p}_i \times \left(\vec{e}_j \times \vec{\ell}_{ji}\right)\right)$$

$$c_{ik}^j = \vec{e}_j^T\left(\vec{e}_i \times \left(J_i^* - \left(\frac{1}{2}\operatorname{tr} J_i^*\right)I\right)\vec{e}_k\right) + \vec{e}_j^T\left(\vec{\ell}_{ji} \times \left(\vec{e}_k \times \left(\vec{e}_i \times \vec{p}_i\right)\right)\right) \tag{1.3.21}$$

$$G_i = \vec{e}_i^T \left(\vec{r}_{ii}^* \times \left(m_i^* \vec{g} \right) \right)$$

The iteration of dynamic parameters is referred to the link coordinate frame.

Let us introduce some symbols:

^iJ_i — is the inertia tensor of link i corresponding to origin O_i referred to link frame i

$^i\bar{J}_i$ — is the inertia tensor of augmented link i corresponding to origin O_i referred to link frame i

$$^i\bar{J}_i = {}^iJ_i + \left(\ell_{i,i+1}^2 I - {}^i\vec{\ell}_{i,i+1} \cdot {}^i\vec{\ell}_{i,i+1}^T \right) m_{i+1}^* , \qquad i=1,\ldots,n-1$$

$$^n\bar{J}_n = {}^nJ_n$$

$$\nu_i = \frac{1}{2} \operatorname{tr} J_i^* \qquad i=1,\ldots,n$$

$$\bar{\nu}_i = \frac{1}{2} \operatorname{tr} \bar{J}_i \qquad i=1,\ldots,n$$

$$^i\bar{P}_i = m_i {}^i\vec{r}_{ii} + m_{i+1}^* {}^i\vec{\ell}_{i,i+1}$$

$$\vec{b}_{ji} = \vec{e}_j \times \vec{\ell}_{ji}$$

(1.3.22)

Note that $^i\bar{J}_i$, $^i\vec{P}_i$ and $\bar{\nu}_i$ are independent of the position of manipulator. In other words, these values can be calculated in advance.

The procedure for computing all elements of H, C and G matrices is follows:

I. $\quad {}^{i+1}\vec{e}_j = A_{i+1}^T {}^i\vec{e}_j \qquad \begin{aligned} j&=1,\ldots,n-1 \\ i&=j,\ldots,n-1 \end{aligned}$

II. $\quad {}^n\vec{P}_n = {}^n\vec{P}_n , \qquad {}^{i-1}\vec{P}_i = A_i {}^i\vec{P}_i$

$\qquad {}^{i-1}\vec{P}_{i-1} = {}^{i-1}\vec{P}_i + {}^{i-1}\vec{P}_{i-1} \qquad i=n,\ldots,2$

III. $\quad {}^i\vec{b}_{ji} = A_i^T \left({}^{i-1}\vec{b}_{i,i-1} + {}^{i-1}\vec{e}_j \times {}^{i-1}\vec{\ell}_{i-1,i} \right) , \qquad {}^i\vec{b}_{jj}=0 \qquad \begin{aligned} j&=1,\ldots,n-1 \\ i&=j+1,\ldots,n \end{aligned}$

IV. $\quad {}^iJ_i^* = A_{i+1} {}^{i+1}J_{i+1}^* A_{i+1}^T + \left(2 {}^i\vec{\ell}_{i,i+1}^T {}^i\vec{P}_{i+1} \right) I -$

$\qquad - {}^i\vec{\ell}_{i,i+1} {}^i\vec{P}_{i+1}^T - {}^i\vec{P}_{i+1} {}^i\vec{\ell}_{i,i+1}^T + {}^iJ_i , \qquad {}^nJ_n^* = {}^n\bar{J}_n \qquad i=n-1,\ldots,1$

$$\text{V.} \qquad v_i = v_{i+1} + 2\, {}^i\vec{\ell}_{i,i+1}^{\,T}\, {}^i\vec{p}_{i+1} + \bar{v}_i, \qquad i=n-1,\ldots,1$$

$$\text{VI.} \qquad H_{ij} = {}^i\vec{e}_i^{\,T}\, {}^iJ_i^*\, {}^i\vec{e}_j + {}^i\vec{e}_i^{\,T}\left({}^i\vec{p}_i \times {}^i\vec{b}_{ji}\right), \qquad \begin{array}{l} i=n-1,\ldots,1 \\ j=2,\ldots,i-1 \end{array}$$

$$H_{ij} = {}^i\vec{e}_i^{\,T}\, {}^iJ_i^*\, {}^i\vec{e}_i \qquad\qquad i=1,\ldots,n$$

$$\text{VII.} \qquad c_{ik}^j = {}^i\vec{e}_k \cdot \left(\left({}^iJ_i^* - v_i I\right)\left({}^i\vec{e}_j \times {}^i\vec{e}_i\right) + \left({}^i\vec{e}_i \times {}^i\vec{p}_i\right)\times {}^i\vec{b}_{ji}\right) \qquad \begin{array}{l} i=2,\ldots,n \\ j=1,\ldots,i-1 \\ k=1,\ldots,i \end{array}$$

$$\text{VIII.} \quad G_i = {}^i\vec{e}_i^{\,T}\left({}^i\vec{p}_i \times {}^i\vec{g}\right)$$

$$^i\vec{g} = A_i^T\, {}^{i-1}\vec{g} \qquad\qquad i=1,\ldots,n \qquad\qquad (1.3.23)$$

When the Denavit-Hartenberg notation is used [69], then

$$A_i = \begin{bmatrix} 1 & 0 & 0 \\ 0 & \cos\alpha_i & -\sin\alpha_i \\ 0 & \sin\alpha_i & \cos\alpha_i \end{bmatrix} \begin{bmatrix} \cos q_i & -\sin q_i & 0 \\ \sin q_i & \cos q_i & 0 \\ 0 & 0 & 1 \end{bmatrix} \qquad (1.3.24)$$

$$\text{and} \quad {}^i\vec{\ell}_{i,i+1} = \begin{bmatrix} a_i \\ 0 \\ d_i \end{bmatrix}, \quad {}^i\vec{e}_i = \begin{bmatrix} 0 \\ 0 \\ 1 \end{bmatrix}, \quad {}^i\vec{b}_{i,i+1} = \begin{bmatrix} 0 \\ a_i \\ 0 \end{bmatrix} \qquad (1.3.25)$$

Number of floating-point multiplications n_M and additions n_A for the computation of H, C, G matrices as well as driving torques P_i, is given by

$$n_M = \frac{3}{2} n^3 + \frac{35}{2} n^2 + 9n - 16$$
$$(1.3.26)$$
$$n_A = \frac{7}{6} n^3 + \frac{23}{2} n^2 + \frac{64}{3} n - 28$$

where n is the number of degrees of freedom. For a manipulator with 6 revolute joints, we have $n_M = 992$ and $n_A = 776$. This is significantly less (even 3 times) than in other numerical methods. More details about this method can be found in [68].

1.4. Methods Based on Appel's Equations

The method of Appel's equations, conceived by E. P. Popov and associates [47], was developed in its final form in papers by M. Vukobratović and V. Potkonjak [48, 26]. It solves both inverse and direct dynamic problem. These equations are based on Gibbs' "acceleration energy" function. In this method, disposition of the links and their kinematics are described in the same way as in Newton-Euler's method.

Relative position of adjacent links is described by the Rodrigues' formula. Transformation matrices A_i^{i-1} between coordinate system of the (i-1)-st link and the local system of the i-th link are obtained in this way. The kinematic variables, such as angular velocity $\tilde{\omega}_i$, angular acceleration $\tilde{\varepsilon}_i$, and linear acceleration $\tilde{w}_i$, are determined with respect to the local coordinate system of the i-th link. The relations determining these kinematic quantities are analogous to relations (1.3.8) from Newton-Euler's method:

$$\tilde{\omega}_i = A_i^{i-1}\tilde{\omega}_{i-1} + \dot{q}_i\tilde{e}_i$$

$$\tilde{\varepsilon}_i = A_i^{i-1}\tilde{\varepsilon}_{i-1} + \ddot{q}_i\tilde{e}_i + \dot{q}_i\left(\tilde{\omega}_i \times \tilde{e}_i\right) \qquad (1.4.1)$$

$$\tilde{w}_i = A_i^{i-1}\left(\tilde{w}_{i-1} - \tilde{\varepsilon}_{i-1} \times \tilde{r}_{i-1,i} - \tilde{\omega}_{i-1} \times \left(\tilde{\omega}_{i-1} \times \tilde{r}_{i-1,i}\right)\right) +$$

$$+ \tilde{\varepsilon}_i \times \tilde{r}_{ii} + \tilde{\omega}_i \times \left(\tilde{\omega}_i \times \tilde{r}_{ii}\right)$$

where: $\tilde{e}_i$ - is the unit vector of the i-th joint axis expressed with respect to the local coordinate system of the i-th link,

$\tilde{r}_{ij}$ - distance vector from the j-th joint to the i-th link mass center, expressed with respect to the local coordinate system of the i-th link.

Mechanism dynamics is described by Gibbs-Appel's equations

$$\frac{\partial G}{\partial \ddot{q}_i} = Q_i, \qquad i=1,\ldots,n \qquad (1.4.2)$$

where: G - is "acceleration energy" function, and

Q_i - is generalized force expressed with respect to the i-th joint.

The "acceleration energy" function G may be expressed by the sum

$$G = \sum_{i=1}^{n} G_i \qquad (1.4.3)$$

where G_i is Gibbs' function relating to the i-th link, which is given by the expression

$$G_i = \frac{1}{2} m_i \tilde{w}_i^2 + \frac{1}{2} \tilde{\varepsilon}_i \cdot \tilde{J}_i \tilde{\varepsilon}_i + 2\left(\tilde{\omega}_i \times \tilde{J}_i \tilde{\omega}_i\right) \cdot \tilde{\varepsilon}_i. \qquad (1.4.4)$$

Substituting (1.4.1) into (1.4.4) and (1.4.2), the dynamic robot model is obtained in the form

$$H(q)\ddot{q} + h_c(q, \dot{q}) = Q \qquad (1.4.5)$$

where: $Q = \begin{bmatrix} Q_1 & \cdots & Q_n \end{bmatrix}^T$ - is the vector of generalized forces, and $H(q)$ and $h_c(q, \dot{q})$ - are $n \times n$ and $n \times 1$ model matrices, respectively. The vector of generalized forces Q may be expressed as the difference between the vector of driving forces (torques) and the vector describing gravity effects $g(q)$:

$$Q = P - g(q).$$

Substituting into (1.4.5), the model is obtained in the form

$$H(q)\ddot{q} + h(q, \dot{q}) = P$$

where $h(q, \dot{q}) = h_c(q, \dot{q}) + g(q)$. It should be stated that these vectors have already been defined when presenting Lagrange's and Newton-Euler's methods. For example, in Uicker-Kahn's method, in model (1.2.6), the vector $h_c(q, \dot{q})$ has been denoted as

$$\dot{q}^T C(q) \dot{q}$$

and represents the vector of Coriolis' and centrifugal effects.

It may be concluded that Appel's method belongs to the group of methods enabling both the direct and the inverse dynamic problem solutions. The number of numerical operations required to form the dynamic model applying Appel's method is

$$n_M = \frac{7}{3} n^3 + 27n^2 + \frac{722}{3} n + 9$$

$$n_A = \frac{10}{3} n^3 + \frac{43}{2} n^2 + \frac{931}{6} n + 6$$

where n_M is the number of floating-point multiplications, and n_A - the number of floating-point additions. For manipulators with 3 and 6 degrees of freedom, we obtain the following numbers of operations

n	n_M	n_A
3	1037	755
6	2929	2431

In the algorithms presented in the previous sections we have accepted that one floating-point multiplication and one addition take about 0.1 ms. We have assumed that algorithms are implemented on contemporary microcomputers equiped with floating-point coprocessors or appropriate hardware support. We have also concluded that the algorithm may contain up to 200 multiplications and additions for a sampling period of 20 ms. Since the algorithm based on Appel's equations requires more than 5 times larger number of operations (for manipulators with 3 revolute degrees of freedom), this algorithm is also inconvenient for real time implementation.

1.5. Symbolic Methods

In the foregoing section we have presented numerical methods for robot mechanisms modelling. In these methods all variables are treated as real numbers. This means that one memory location is assigned to each variable in program implementation. The value of the variable is determined by the contents of the appropriate memory location. In this and in the next section we will describe analytical methods for robot modelling. In these methods a variable is represented by a sets of symbols which describe the structure of the variable, i.e. its dependence on parameters, joint coordinates and joint rates. In the text to follow the analytical expressions will be referred to as "symbolic expressions" or "symbolic forms".

The beginning of the last decade has seen the development of a larger number of software packages handling symbolic algebraic and transcedental expressions. These are, for example, programs for algebra of functions [49], for operation with rational functions [50, 51], etc. These software packages have a similar structure:

1. input module for entering symbolic expressions together with an editor,

2. module for syntactic analysis and identification of expressions,

3. module for translating symbols into the internal code,

4. module for algebraic manipulations and simplifications (which should imitate appropriate human operation), and

5. output module.

Evidently, these packages were very complex because of the extreme diversity of the expressions operated with in the applied physics and mathematics, and their implementation required very high performance computers. The computer time required to generate complex symbolic forms figuring in nonlinear mathematical models was especially critical. This is why packages oriented to certain fields of applied physics and chemistry were also developed. For example, such a package was developed for use in theoretical seismology [52]. However, even the special-purpose programs were inapplicable to more complex problems of a particular engineering field since the mathematical models are oftenly, very complex and nonlinear. To illustrate this let us mention the class of spatial mechanisms, i.e., systems of rigid or elastic bodies interconnected by joints, such as robots, satellites, rehabilitation devices, electro-hydraulic systems, etc. With such systems there exists a dynamic interaction (coupling) between links due to inertial, Coriolis', centrifugal and external forces and moments. The interaction forces depend on the instantaneous disposition of the links, velocity and acceleration. These forces act between all links, i.e., the motion of a single link affects, as a rule, the motion of the remaining ones. Thus, it follows that the complexity of mathematical functions which describe robot dynamics increases rapidly with the increase in the number of links, i.e., the number of the degrees of freedom. Because of such model complexity, practically none of the software packages for symbolic algebra could properly handle robot dynamic modelling. Some attempts were made [53] to develop specific software packages for analyzing open spatial mechanisms. However, they appeared to be practically applicable to systems with no more than 3 degrees of freedom[*).

[*) Ref. [53] presents an example of a 4 dof manipulator where axes of even 3 joints are parallel resulting in considerably simplified model.

1.6. Numeric-Symbolic Method

In this section we will present the fundamentals of a new computer aided method for robot modelling which we have termed "numeric-symbolic" method. The method will be described in detail in the following chapters.

In the previous section we have described the difficulties arising in computer construction of dynamic robot equations in analytical, i.e. symbolic form. These difficulties were the reason for which the idea about symbolic modelling was practically abandoned in the past ten years and, with a number of exceptions, was not considered in the literature [54].

On the other hand, it was just during this past decade that numerical methods for robot modelling were intensively developed. In the brief review we have presented, one could distinguish between two classes of these methods: 1. methods yielding the dynamic model matrices, i.e. enabling solution to both direct and inverse problem of dynamics, and 2. methods solving the inverse problem only, i.e. computing driving forces (torques) given a manipulator motion. In addition, it was concluded that the methods belonging to class 1. are hardly implementable in real time because of the large number of numerical operations. The algorithms from class 2. contain considerably fewer numerical operations, but they are also inconvenient for real time implementation for manipulators with 3 and more degrees of freedom. These difficulties gave rise to the development of multiprocessing algorithms [55], which are still being elaborated at the level of simulation because of the complexity and high implementation cost. It should be stated that the algorithms from class 2. cover only a class of control algorithms which are based on direct calculation of driving forces. These methods are inapplicable to control laws employing dynamic model matrices and to computer simulation of robot motion. Thus, it may be concluded that numerical methods also do not give satisfactory results, especially as far as their implementation on today's general-purpose microcomputers is concerned.

Analyzing the numerical methods, we can conclude that the large number of numerical operations results from the fact that the numerical methods inherently incorporate all kinematic and dynamic laws which are employed in mathematical model construction. For example, Newton-Euler's method [20 - 22] incorporates the Rodrigues' formula, the laws of the linear and angular velocity of a rigid body, Euler's dynamic equations, and the conditions for kinetostatic equilibrium. "Passing through" these laws during model computation requires, however, a long processing time and is, therefore, hardly implementable in real time. One may now raise the question of relationship between the number of numerical operations required to calculate the model matrices in symbolic form and the number needed in the corresponding numerical algorithm (which gives the same model). Since the symbolic model does not require "passing through" the laws of mechanics, but represents a set of analytical expressions for direct calculation of the elements of model matrices, it is obvious that the symbolic model contains considerably less number of numerical operations than the numerical algorithms.

However, as shown in the preceding text, the problem of computer-aided generation of symbolic model is a very complex problem. This is why it has not been solved in a satisfactory manner so far. To explain the new modelling concept which will be described in detail and verified in the following text, let us consider the essential differences between symbolic and numerical approach to robot modelling. As may be seen from Table 1.1[*)], both modelling concepts have their advantages as well as

APPROACH PROPERTY	SYMBOLIC	NUMERICAL
Complexity of the modelling algorithm	extremely high	low
Number of numerical operations for model matrices calculation	small	large

Table 1.1. Comparison of properties of symbolic and numeric methods

disadvantages. One may now ask whether "hybridization" of these two approaches may yield a new concept which will incorporate the advantages of both approaches. In the further text we will develop a concept having just these properties and will call it "numeric-symbolic" or "analytic-

---

[*)] The properties of modelling given in this table are relative, so "extremely high" means "extremely high complexity of the algorithm by which the symbolic model is formed compared with numerical approach", and so on.

al" method for robot modelling. In the text which follows immediately
we will present only the basic ideas underlying this method, while the
complete algorithm will be described in the following chapters.

Let us consider a manipulator with n links and n joints, so that the
whole mechanism has n degrees of freedom in total. Each link is charac-
terized by a set of parameters. When presenting Uicker-Kahn's method,
we have shown that transformation matrices are described by 3 geomet-
rical parameters $\left(a_k,\ \alpha_k,\ s_k\right)$. A complete definition of each link also
requires specification of mass and inertia tensor as well as the posi-
tion of mass center. Accordingly, each link is described by 16 parameters. It is the same case with other methods. For example, the methods
employing Rodrigues' formula describe a link by 12 geometrical parameters, mass and 3 principal moments of inertia. Thus, we conclude that
16n parameters figure in the mathematical model of a robot with n degrees of freedom. When symbolic modelling is considered, these 16n parameters are treated as 16n symbols. Adding n joint coordinates and n
joint velocities, we obtain a total of 18n symbols. It is now evident
that manipulation with functions depending on 18n variables is extremely complex. For a 6-degree-of-freedom manipulator, over 100 symbolic
variables figure in the symbolic expressions, which include sines and
cosines of joint angles. However, this is just one among the reasons
why software packages for handling symbolic expressions were not capa-
ble of generating the robot model. Another reason is the extreme complexity of the expressions themselves. This is why we will develop the
numerical-symbolic method in such a way that these expressions become
as simple as possible. Therefore, we will accept the following premises
[56 - 58]:

1. parameters will be treated as numerical constants, and numerical operations between them as operations between numerical
   values, and

2. symbolic expressions will be treated as functions of joint coordinates only.

Evidently, the first assumption eliminates the 16n symbolic variables
corresponding to system parameters. To satisfy the second premise, we
will present the robot model in the form

$$P_i = \sum_{k=1}^{n} H_{ik}(q)\ddot{q}_k + \sum_{k=1}^{n}\sum_{\ell=1}^{n} C_{k\ell}^{i}(q)\dot{q}_k\dot{q}_\ell + h_G(q) \qquad (1.6.1)$$

44

for i=1,...,n. This model form is easily recognizable in Uicker-Kahn's
and in many other methods. Eq. (1.6.1) may be written in the following
matrix form

$$P = H(q)\ddot{q} + \dot{q}^T C(q)\dot{q} + h_G(q) \qquad (1.6.2)$$

where P - is n-dimensional vector of driving torques (forces),

   q(t) - n-dimensional vector of joint coordinates,

   H(q) - n×n inertial matrix,

   C(q) - n×n×n matrix of Coriolis' and centrifugal effects, and

   $h_G(q)$ - is n-dimensional vector of gravity effects.

Here $\dot{q}^T C(q)\dot{q}$ denotes the vector

$$\dot{q}^T C(q)\dot{q} = \begin{bmatrix} \dot{q}^T C^1(q)\dot{q} \\ \vdots \\ \dot{q}^T C^n(q)\dot{q} \end{bmatrix} \qquad (1.6.3)$$

where $C^i(q) = \left[C^i_{k\ell}(q) \right]$ is n×n matrix.

One may grasp the sense of the second premise from the form of the given
dynamic model. The idea is to manipulate with symbolic expressions which
depend on joint coordinates only. Since joint velocities figure in the
model in a regular quadratic form of $\dot{q}^T C^i(q)\dot{q}$ type, there is no reason
for treating the joint velocities as symbols in process of symbolic
model generation.

We may now derive the following conclusions: the matrices H(q), C(q)
and the vector $h_G(q)$ depend on joint coordinates only and are independent of joint velocities. By determining these matrices we determine
completely the mathematical robot model. Since these matrices are functions of joint coordinates only, premise 2. may be satisfied, because
the elements of these matrices may be treated as symbolic expressions
depending on joint coordinates only. In the following text these matrices will be referred to as "dynamic model matrices" or, shorter, "dynamic matrices".

Let us now briefly consider the problem of forming the dynamic model
matrices. It follows from the review of numerical methods that these

matrices may be formed by Uicker-Kahn's method [1 - 5], algorithms by S. Mahil [12] and M. Renaud [20], which are based on Lagrange's equations, or by Huston-Kane's algorithm [43 - 45] based on Newton-Euler's equations. However, none of these algorithms gives the elements of dynamic matrices in a suitable form. For example, the partial derivatives of transformation matrices and inertia tensors figure in Uicker-Kahn's method. The inertia tensor also figures in Renaud's algorithm. Evidently, it would be much more convenient to use 3 principal moments of inertia instead of the inertia tensor. Expressions for dynamic matrices given in the algorithms by S. Mahil and R. Huston are not simple enough either. This is why we will develop in the following chapter the expressions which provide the simplest possible description of the elements of dynamic matrices. This derivation will be based on Newton-Euler's method [20 - 22]. Since this method is recursive and does not give the elements of these matrices in a closed form, we will preformulate it. As a result, we will obtain very compact expressions for the elements of dynamic matrices. For example, the (i, k)-th element of inertial matrix is obtained in the form[*)]

$$H_{ik} = \sum_{j=\max(i,k)}^{n} \left(m_j \left(\vec{e}_i \times \vec{r}_{ji} \right) \cdot \left(\vec{e}_k \times \vec{r}_{jk} \right) + \vec{e}_i J_j \vec{e}_k \right) \qquad (1.6.4)$$

where: m_i - is mass of the i-th link,

$\quad J_i$ - inertial matrix with respect to the fixed system,

$\quad \vec{e}_i$ - unit vector of the i-th joint axis in the reference system,

$\quad \vec{r}_{ji}$ - distance vector from the i-th joint to the j-th link mass center, expressed in the reference coordinate system.

It should be noted that the structure of expression (1.6.4) is much simpler than that of (1.3.6), the corresponding expression from Newton-Euler's method. Namely, expression (1.3.6) incorporates coefficients which are determined by a series of recursive relations from the kinematic and dynamic stages of modelling (1.3.1 - 1.3.5).

The elements of the remaining dynamic matrices are obtained similarly to expression (1.6.4). On the basis of these expressions, it is easy to prove some properties of model matrices, such as symmetry and positive definiteness of inertial matrix, symmetry and antisymmetry of matrices $c^i(q)$, etc. In addition, it is easy to derive the linearized robot

---

[*)] This expression holds for a manipulator with revolute joints. When there exist sliding joints, the expression is simplified (see Ch. 2).

model as well as the models of sensitivity to variations in dynamic
parameters. These algorithms will be presented in detail in Para. 2.4.
and 2.5.

Let us now return to the problem of forming the symbolic expressions.
To give a simple explanation of the basic idea of numerical-symbolic
method, we will consider a simple example. The analytical model of an
arthropoid robot with 3 degrees of freedom will be derived at the end
of this book, in Para. 5.2. It will be shown that the element $H_{11}(q)$ of
the inertial matrix for this mechanism has the following analytical
form

$$H_{11}(q) = 4.57 + 112.98 \sin^2 q_2 + 26.8 \cos q_3 \sin^2 q_2 +$$

$$+ 26.8 \cos q_2 \sin q_2 \sin q_3 + 2.68 \cos^2 q_3 \sin^2 q_3 \quad (1.6.5)$$

$$+ 5.36 \cos q_2 \cos q_3 \sin q_2 \sin q_3 + 2.68 \cos^2 q_2 \sin^2 q_3$$

This expression may be represented in a unique manner by an arranged
set of numerical data:

    1. by the vector of coefficients, and

    2. by the matrix of exponents,

i.e. by the following arranged pair

$$
\begin{bmatrix}
4.57 \\
112.98 \\
26.80 \\
26.80 \\
2.68 \\
5.36 \\
2.68
\end{bmatrix}
,
\begin{bmatrix}
0\ 0\ 0 & 0\ 0\ 0 & 0\ 0\ 0 \\
0\ 0\ 0 & 0\ 2\ 0 & 0\ 0\ 0 \\
0\ 0\ 1 & 0\ 2\ 0 & 0\ 0\ 0 \\
0\ 1\ 0 & 0\ 1\ 1 & 0\ 0\ 0 \\
0\ 0\ 2 & 0\ 2\ 0 & 0\ 0\ 0 \\
0\ 1\ 1 & 0\ 1\ 1 & 0\ 0\ 0 \\
0\ 2\ 0 & 0\ 0\ 2 & 0\ 0\ 0
\end{bmatrix}
\qquad (1.6.6)
$$

Correlation between this set of data and the analytical expression is
now easy to establish. The elements of the vector of coefficients directly correspond to the coefficients of analytical expression. One
row of the matrix of exponents corresponds to each coefficient. This
row describes the contents of the analytical expression accompaning the

corresponding coefficient. Thus, the row

$$\left[c_1 \cdots c_n \quad s_1 \cdots s_n \quad u_1 \cdots u_n \right]$$

has its corresponding in the analytical expression

$$\left(\cos q_1 \right)^{c_1} \cdots \left(\cos q_n \right)^{c_n} \left(\sin q_1 \right)^{s_1} \cdots \left(\sin q_n \right)^{s_n} q_1^{u_1} \cdots q_n^{u_n}$$

where c_i, s_i and u_i are integer exponents and $\cdot$ stands for multiplication. Let us consider the example of the third row of the vector of coefficients and the matrix of exponents (1.6.6):

$$([26.80], \; [0 \; 0 \; 1 \quad 0 \; 2 \; 0 \quad 0 \; 0 \; 0]).$$

The third addend in analytical expression (1.6.5) corresponds to that row:

$$26.80 \cos q_3 \sin^2 q_2.$$

Evidently, the number of rows in (1.6.6) corresponds to the number of addends in analytical expression (1.6.5).

The above-defined, arranged structure of numerical data will be referred to as "the polynomial matrix". Any variable participating in the construction of mathematical model may be described by such polynomial matrices. In Ch. 2 we will prove a very interesting property of the matrix of exponents corresponding to any model variable: the elements of the matrix of exponents can only have the values 0, 1 or 2. This property is particularly important for program implementation of the algorithm. It is now evident that, if we wish to substitute the variables (participating in forming the model matrices) by polynomial matrices, we should develop the appropriate algebra for polynomial matrices. As will be shown in Para. 3.2, the algebraic operations between polynomial matrices reduce to simple operations between the vectors of coefficients and matrices of exponents, and these operations are very suitable for programming. These results were used to develop numerical--symbolic algorithms for robot modelling, such as

1. algorithm for generating dynamic robot model (Para. 3.4),

2. algorithm for generating linearized dynamic model (Para. 3.5),

3. algorithm for generating model of sensitivity to variations in

dynamic parameters (Para. 3.5), and

4. algorithm for deriving approximate models, given an admissible relative error (Para. 3.6).

Further problem which arises is the problem of optimal computation of analytical expressions which correspond to the obtained polynomial matrices. For the sake of clarity, let us return to the previous example. The analytical form of element $H_{11}(q)$ of the arthropoid robot contains 20 multiplications and 6 additions, as may be seen by direct enumeration of the operations. On the other hand, the same form may be represented in factorized form

$$H_{11}(q) = \sin q_2\Big(\sin q_2\Big(112.98 + \cos q_3\big(26.80 + 2.68 \cos q_3\big)\Big) +$$
$$+ \sin q_3\cos q_2\big(26.8 + 5.36 \cos q_3\big)\Big) + \qquad (1.6.7)$$
$$+ 2.68 \sin^2 q_3\cos^2 q_2 + 4.57.$$

In this form there are 11 multiplications and 6 additions. This means that we have saved nearly 50% of the number of multiplications. The problem of the optimal factorization of numerical-symbolic forms (in the sense of a minimal number of numerical operations) is treated in detail in Ch. 4. It will be shown that the structure of optimal calculation may be represented by a graph or appropriate matrices corresponding to the graph. For example, the previously derived expression for $H_{11}(q)$ may be represented by the graph shown in Fig. 1.5, where nodes correspond to the elements of expression $H_{11}(q)$, and branches to the multiplications of these elements. The elements are represented in a shortened form as $c_2 = \cos q_2$, $c_3 = \cos q_3$, $s_2 = \sin q_2$ and $s_3 = \sin q_3$ and constants as $K1 = 4.57$, $K2 = 112.98,\ldots,K7 = 2.68$ (in accordance with the order in the vector of coefficients). The operation of addition is described by the flow of several branches into one node. Such a graph may be described by two matrices, one of them containing the elements in nodes and the other describing the branching structure. Thus, the problem formulated reduces to determining "the matrix representatives of the sequences of optimal calculation". When these matrices have been determined for all analytical expressions which participate in constructing the mathematical model, the problem arises how to generate "a program for real-time operation" on the basis of these matrices. It will be shown in Ch. 4 that the matrix representative of the graph may be used to generate a series of recursive relations for cal-

culating the corresponding analytical expression. The program which
forms such recursive relations will be referred to as "expert-program".
The output of this program should represent a series of program instructions by which the model is formed. Evidently, the expert-program
may be realized so that the output program code is written in the language used by the microcomputer on which the model is to be implemented
or in a higher-level programming language.

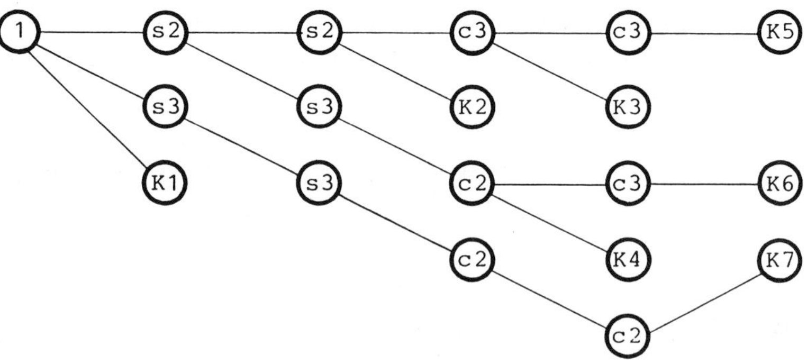

Fig. 1.5. Structure of optimal calculation

Fig. 1.6. shows a general block-diagram of numerical-symbolic method
with all stages of program generation for real-time operation. This
block-diagram presents, in fact, all basic components of the software
package whose input data are the robot parameters and output represents
the source program of robot model. The software package is organized in
several parts:

1. Input of geometric and dynamic parameters of robot mechanism by
   an interactive procedure;

2. Choosing the model type - nonlinear, linearized, approximate,
   or sensitivity model. Then, the numeric-symbolic model is formed by using a program for the algebra of polynomial matrices;

3. On the basis of the obtained analytical forms optimization of
   their calculation is performed;

4. Expert-program is used to generate the program code (in the
   language of a desired microcomputer) which is suitable for
   real-time execution.

Let us note that this software package is to be implemented on medium-
-class computers (e.g., minicomputers). However, the output program is

most oftenly intended to be executed on microcomputer in the robot control system.

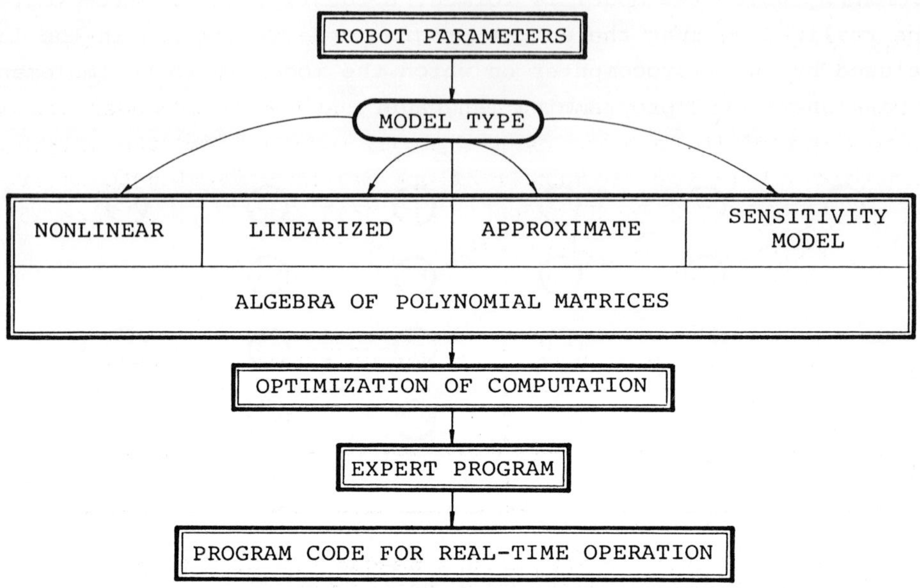

Fig. 1.6. General block-diagram of the algorithm for generating
robot models for real time operation

All previously presented results will be illustrated by examples of
actual industrial robots in Ch. 5. Robots of cylindrical, arthropoid
and anthropomorphic structures have been chosen to illustrate the dependence of mathematical model on manipulator structure. These examples
are used to illustrate methods for forming nonlinear, linearized and
sensitivity models in numeric-symbolic form. The program code generated by "expert program" will be presented at the end of each example.
The number of multiplications and additions required to form the models
will be derived therefrom.

Analysis of the results obtained will show that the number of numerical
operations in the models generated is even several times smaller
than the corresponding number in numerical methods. To illustrate the
extremely high savings in the number of operations, we will here com-
pare one of the numerical methods (method in 1.3.3) with the developed
numerical-symbolic method. The results of this comparison are given in
Table 1.2. The anthropomorphic robot whose model is derived in Para.5.3

(Table 5.39) has been chosen as example. The number of operations according to the recursive numeric method have been obtained from the expressions

$$n_M = \frac{3}{2} n^3 + \frac{35}{2} n^2 + 9n - 16 \quad \text{and} \quad n_A = \frac{7}{6} n^3 + \frac{23}{2} n^2 + \frac{64}{3} n - 28$$

presented in 1.3.3. These expressions give the number of numerical operations required to form the dynamic model matrices and the driving torque vector. Data for the number of operations according to numeric symbolic method have been obtained in the following way. In Para. 5.3 it will be shown that the computation

	New recursive method	Numerical-symbolic method
n_M	209	65
n_A	171	29

Table 1.2. The number of operations needed for computing model matrices for anthropomorphic robot

of dynamic model matrices (1.6.2) for anthropomorphic robot requires n_M = 44 floating-point multiplications and n_A = 23 additions. In order to ensure a fair comparison of these methods, one should add to these numbers 21 multiplications and 6 additions required to calculate the vector $h(q, \dot{q}) = \dot{q}^T C(q) \dot{q} + g(q)$. Thus, one obtains n_M = 65 and n_A = 29 from Table 1.2. As may be seen from this table, the number of multiplications in reduced 3.2 times and the number of additions 5.9 times. Assuming that one multiplication lasts as long as two additions, we obtain that the number of operations is effectively smaller about 4 times.

In presenting the methods in preceding paragraphs we have accepted that one multiplication and one addition take about 0.1 ms on up-to-date microcomputers. We have also seen that none of the numerical methods can be implemented in real time (which means that the model cannot be formed in 20 ms) for manipulators with 3 and more revolute degrees of freedom. However, Table 1.2 shows that the numeric-symbolic model of antropomorphic manipulator can be formed in about 5 ms, which represents only a quarter of the maximal acceptable time. This, in turn, means that not only the models of manipulators with 3 but also with 4, and sometimes even with 5 and 6 degrees of freedom can be formed in real time. Of course, all these conclusions apply to a robot control system with a single processor. The implementation of numeric-symbolic method on multiprocessor systems represents an important problem which is out of the scope of this book.

Although the programs for real-time operations based on numeric-symbolic method have considerably better performances than classical, numerical programs, this new approach has a constraint which should be pointed out here. If a kinematic parameter is changed, it is necessary to generate a new program for real-time operation using the developed software package. This is not the case with classical methods. However, if a dynamic parameter is changed, it is sufficient to change the constants in the existing program in accordance with the sensitivity model (Ch. 3), and no new program code should be generated.

Chapter 2
Computer-Aided Method for Closed-Form Dynamic Robot Model Construction

2.1. Introduction

A new method for modelling of robotic system will be presented in this chapter. In addition, some fundamental properties of dynamic model matrices will be proved.

For a precise derivation of the mathematical model, basic notions relating to the physical model of robot mechanism will be defined at the beginning of this chapter. Definitions of configuration, link (segment), kinematic pair, kinematic chain, joint and external coordinates, and active mechanism will be provided. The main assumptions which will be accepted for the robot physical model may be reduced to the following three:

1) links are modelled by rigid bodies, and the effects of elasticity are neglected;

2) the kinematic chain of robotic manipulator is neither branched nor closed;

3) there are no kinematically coupled degrees of freedom.

These assumptions are mainly satisfied for the class of manipulation robots, while they need not hold for locomotion robots, complex hydraulic and other mechanical systems. This is why "a robot" should be understood as a manipulation robot in the text to follow.

One of the well-known methods for mathematical modelling of manipulators, based on D'Alemberts principle and Newton-Euler equations [20 - 22], will be presented in Para. 2.2. It will be seen that this procedure consists of the following stages: 1) determination of the spatial disposition of robot links with respect to a reference coordinate system, 2) kinematic stage, and 3) dynamic stage. A series of geometric, kinematic and dynamic variables and parameters connected by recursive relations appear in this method. These variables are first calculated for the first link, then for the second using previous results etc.

Analyzing the algorithm of Newton-Euler method (see 1.3.1), one may note that the series of recursive relations in this method represents in fact a series of laws, i.e. theorems of kinematics and dynamics of a system of interconnected rigid bodies. Therefore, one may ask an important question about the very essence of such an approach. Is the mathematical model obtainable in a closed form, without using a series of recursive relations? This would mean a direct generation of differential equations of the model based on known geometric and dynamic parameters, without passing through all laws of mechanics. This problem will be considered in Para. 2.3. Here, the problem of closed-form model construction will be precisely formulated with the following requirements:

1) The influence of inertial effects should be isolated in the model by forming the inertial matrix of the system;

2) The influence of Coriolis and centrifugal effects should be isolated by a separate set of matrices whose elements depend on the joint coordinates only and are independent of joint velocities;

3) The influence of gravity effects should be isolated by a separate column - matrix;

4) The elements of these matrices should be expressed in terms of the simplest expressions without using recursive relations.

Let us note immediately that conditions 1), 2), and 3) require all model matrices to depend on the joint coordinates only, i.e. the joint velocities and accelerations to figure explicitly in the model. As was explained in Ch. 1, this is of particular importance for the development of analytical methods of robot modelling.

The basic Newton-Euler method, briefly presented in 1.2.1, will be used as a basis in starting to solve the problem stated. In deriving the closed-form model, recursive relations will be transformed into non-recursive. Since this procedure results in very complex expressions, it will appear necessary to introduce a series of algebraic identities. Two types of identities will be introduced:

1) physical identities for establishing relations between physical quantities, i.e. kinematic and dynamic variables, and

2) mathematical identities for reducing multiple sums to symmetric
   quadratic forms.

The physical identities, for example, are the identities relating the
joint axes unit vectors to the unit vectors of link coordinate systems.
The idea for obtaining symmetric quadratic forms by mathematical identities lies in the fact that Coriolis and centrifugal effects depend
on the product or square of the joint velocities. That is, Coriolis and
centrifugal forces may be described by a quadratic form which, in a
general case, has the following structure

$$\vec{F}_c = \sum_{k=1}^{n} \sum_{\ell=1}^{n} \vec{f}_c(k, \ell)\dot{q}_k\dot{q}_\ell, \qquad\qquad (2.1.1)$$

where $\dot{q}_i$ represents the joint velocity, $\vec{F}_c$ - the total force, $\vec{f}_c(k, \ell)$
- the component of force due to product $\dot{q}_k\dot{q}_\ell$, and n - the number of
degrees of freedom. This quadratic form may easily be reduced to the
symmetric form

$$\vec{F}_c = \dot{q}^T \phi_c \dot{q} \qquad\qquad (2.1.2)$$

where the symmetric matrix ϕ_c contains elements $\vec{f}_c(k, \ell)$ which depend
on joint coordinates only, and not on joint velocities. Here, one can
recognize requirement 2) imposed on the construction of closed-form
model.

Since the procedure for closed-form model construction is rather complex, its basic idea and the most important relations will be presented in Para. 2.3, while the remaining details of derivations will be
given in Appendices 2.1 and 2.2.

The obtained closed-form model of mechanism will be summarized by a
detailed flow-chart provided at the end of Para. 2.3. It will be seen
that the relations of the obtained model are, in spite of a relatively
complex derivation, very simple and elegant. The matrices of the closed-
-form model will be shown to have the following properties:

1) The elements of model matrices depend on 3 sets of vector variables
   only: unit vectors of joint axes, unit vectors of link coordinate
   systems, and the distance vectors between joints and mass-centers
   of links.

2) Dynamic parameters of links (masses and moments of inertia) linearly

figure in all elements of model matrices;

3) If sliding joints exist, relations of the obtained model become considerably simpler.

A number of another basic properties of dynamic model matrices will
be proven in Para. 2.4. It will be seen that these properties do not
depend on the modelling algorithm but represent fundamental properties
of the mathematical models of open-chain mechanisms:

1) The inertial matrices are symmetric and positive definite.

2) The matrices of Coriolis and centrifugal effects are symmetric. In
   addition, these matrices also possess certain properties of antisymmetry when different degrees of freedom are simultaneously considered.

These properties will be precisely formulated and proven by appropriate theorems. The problem of the complete, nonlinear mathematical model
construction will thus be solved.

A method for the construction of linearized dynamic closed-form model
will be presented in the text to follow. Having in mind that the joint
velocities and accelerations explicitly figure in the derived mathematical model, the problem of determining the linearized model reduces
to determining the matrices of partial derivatives with respect to the
joint coordinates. To determine these matrices, three theorems are
stated and proven in Para. 2.5:

1) Theorem of the partial derivatives of link-attached vectors, i.e.
   the vectors which are invariant with respect to the corresponding
   link coordinate frames.

2) Theorem of the partial derivatives of distance vectors, and

3) Theorem of the partial derivatives of the cross or dot product of
   two arbitrary vectors.

These theorems have been applied to the equations of the closed-form
model to derive the linearized model of the mechanism. At the end, the
model is summarized by a detailed flow-chart.

An algorithm for the construction of closed-form sensitivity model will be presented at the end of this chapter. We shall accept that the sensitivity model (39 - 40) answers the following question: what increment of driving torques (forces) can compensate for a variation of the dynamic parameters of links, so that the joint coordinates and velocities remain unperturbed? By the dynamic parameters we mean the mass and inertias. A variation in parameters is usually characteristic for the last link - the end-effector, because of the variable parameters of workpieces. At the end of deriving it will be whown that the matrices participating in its construction are mathematically and physically analogous to the inertial matrix, the matrix of Coriolis and centrifugal effects and to the gravity vector. Moreover, the properties of symmetry, positive definiteness and antisymmetry also hold for the matrices of sensitivity model. However, the essential difference lies in the fact that the matrices of sensitivity model are independent of dynamic parameters, so they hold for both small and large variations in these parameters. This is a very important property, since finite parameter variations are very usual in robotics. The presented algorithm is summarized by a detailed flow-chart at the end of this chapter.

2.2. Model Based on Newton-Euler's Equations

Let us consider the model of the robot mechanism shown in Fig. 2.1. The model consists of n rigid bodies which represent the links of the mechanism chain. These links are interconnected by prismatic (sliding) or rotational joints. For a precise derivation of differential equations of the robot dynamics, let us introduce a number of basic definitions.

Configuration

Configuration of the mechanism is represented by an arranged n-tuple $\left(J_1, \ldots, J_n \right)$, such that for each $i \in N = \{1, \ldots, n\}$, $J_i \in \{R, T\}$, where R denotes a rotational and T a prismatic joint.

To illustrate this, the configuration RTT-RRR stands for a mechanism with 6 joints, where the second and third joints are sliding, and the remaining ones are rotational.

Fig. 2.1. Model of robot mechanism with n links
and n joints

Link

A link is defined by an arranged set of parameters $C_i\left(K_i,\ \mathcal{D}_i\right)$, where K_i represents a set of kinematic, and $\mathcal{D}_i$ a set of dynamic parameters.

The sets K_i and $\mathcal{D}_i$ may be defined in various ways. In the basic Newton--Euler method, as well as in the method which will be developed in this section, these sets are assumed to have the form

$$K_i = \left(Q_i,\ \tilde{R}_i,\ \tilde{E}_i\right)$$

$$\mathcal{D}_i = \left(m_i,\ \underline{\underline{J}}_i\right)$$

with

$$Q_i = \left(\vec{q}_{i1},\ \vec{q}_{i2},\ \vec{q}_{i3}\right)$$ — the internal (local) ortonormal coordinate system attached to the link i;

$$\tilde{R}_i = \left\{\tilde{r}_{ik}^{*)}\right\}$$ — the set of the distance vectors from points Z_{ik} to the origin of coordinate system Q_i, where the points Z_{ik} represent the centers of joint connections between the i-th and k-th mechanism links, $k\in\{1,\ldots,k_i\}$;

$$\tilde{E}_i = \left\{\tilde{e}_{ik}\right\}$$ — the set of the unit vectors of joint axes by which the i-th link C_i is connected to the remaining links C_k in points Z_{ik};

---

*) ~ denotes the vectors with prescribed projections with respect to the link coordinate system Q_i.

m_i - mass of the i-th link;

$\underline{\underline{J}}_i$ - inertia tensor of the i-th link defined with respect to the local system Q_i.

Frequently we accept that the axes of coordinate system Q_i are aligned along the principal axes of inertia, so the inertia tensor $\underline{\underline{J}}_i$ reduces to three moments of inertia $\underline{\underline{J}}_i = \left(J_{i1}, J_{i2}, J_{i3}\right)$. In addition, the origin of coordinate system Q_i is accepted to coincide with the center of mass of link i. An example of a robot link is shown in Fig. 2.2.

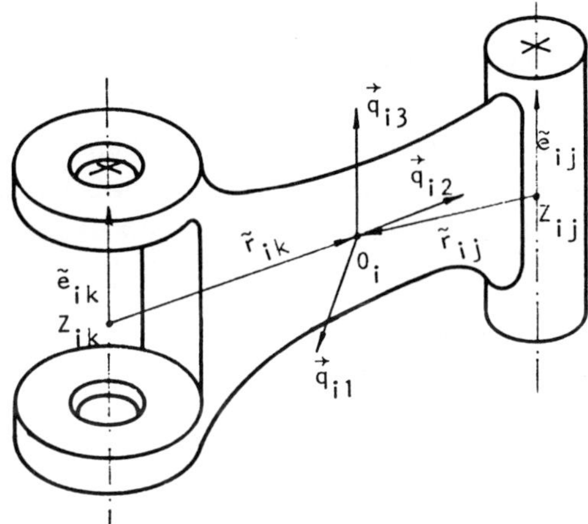

Fig. 2.2. Link C_i with two joints $\left(Z_{ik}$ and $Z_{ij}\right)$ with the center of mass in point O_i

Kinematic pair

A kinematic pair P_{ik} represents a set of 2 adjacent links $\left\{C_i, C_k\right\}$ interconnected by a joint in point Z_{ik}.

The notion of class and subclass of a kinematic pair is introduced depending on the type of joint connection. A j-th class kinematic pair (j=1,...,5) is defined as a set of 2 adjacent links interconnected by a joint with n = 6-j degrees of freedom. A kinematic pair of j-th class and ℓ-th subclass is defined as a pair having r rotational and t prismatic joints in point Z_{ik}, where

$$r = \begin{cases} m-\ell+1 & \ell \leqslant m+1 \\ 0 & \ell > m+1 \end{cases}$$

$$t = s-r.$$

m denotes the maximum possible number of rotational degrees of freedom in the j-th class. For example, classes 1, 2 and 3 permit 3 rotations ($m = 3$), class 4 - two, and class 5 only one rotation.

Kinematic chain

A kinematic chain Λ_n is a set of n interconnected kinematic pairs, $\Lambda_n = \left\{ P_{ik} \right\}$, $i \in N$, $k \in N$.

According to the structure of connections, chains are classified into simple, complex, open and closed.

A chain in which no link C_i enters into more than 2 kinematic pairs is said to be a simple kinematic chain. On the other hand, a complex kinematic chain contains at least one link C_i, $i \in N$ which enters into more than 2 kinematic pairs.

An open kinematic chain possesses at least one link which belongs to one kinematic pair only. If each link enters into at least two kinematic pairs, the chain is said to be closed.

Joint coordinates

Scalar quantities which determine, in a unique manner, the relative position of the links of kinematic pair $P_{ik} = \left\{ C_i, C_k \right\}$ are referred to as manipulator joint coordinates q_{ik}^ℓ. The superscript $\ell \in \{1,\ldots,s\}$, where $s = 6-j$ is the number of degrees of freedom, and j the class of pair P_{ik}.

The definition of joint coordinates may be simplified if we consider fifth-class pairs only, i.e. pairs with a single degree of freedom of relative motion. This assumption does not reduce the generality of consideration, since a pair of any class may be represented by superposition of virtual fifth-class pairs. It is therefore particularly significant to introduce, for such pairs, a unique manner of defining the joint coordinates. Fig. 2.3 shows a revolute kinematic pair whose joint angle (coordinate) is denoted by q_i. A corresponding sliding pair is shown in Fig. 2.4.

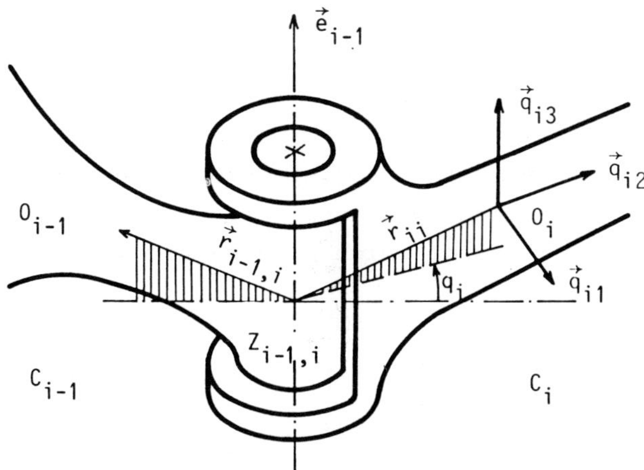

Fig. 2.3. Joint coordinate of a revolute kinematic pair $\left\{C_{i-1},\ C_i\right\}$

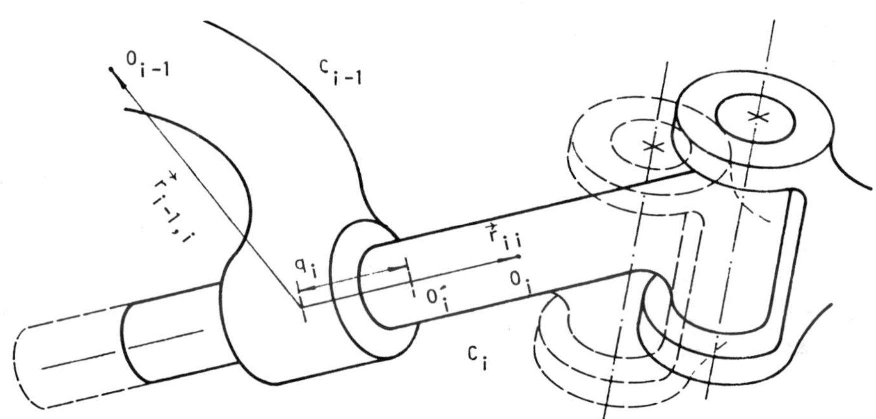

Fig. 2.4. Joint coordinate of a sliding kinematic pair $\left\{C_{i-1},\ C_i\right\}$

External coordinates

Scalar quantities x_{ek}, $k\in\{1,\ldots,m\}$, which determine the position and (partially or completely) orientation of the n-th link of chain Λ_n with respect to a reference coordinate system are said to be the external coordinates of mechanism.

The specification of external coordinates, and of their number $m \leq 6$, depends on the type of manipulation task and will not be considered here.

Active mechanism

An active mechanism represents a system comprising: a) a mechanical part which may be modelled by an appropriate kinematic chain, and b) a set of actuators by means of which driving torques (forces) are realized.

To form the mathematical model of an active mechanism, it is necessary to:

1) identify the parameters of kinematic chain and actuators,

2) form the dynamic model of mechanism, and

3) form the models of actuators.

The dynamic model of open active mechanism represents a set of n nonlinear differential equations which describe system motion in the space of joint coordinates under the action of driving forces.

In the text to follow we will present one of the recursive numerical methods for forming dynamic models of open active mechanisms: the basic Newton-Euler method [20 - 22]. This method consists of the following stages: determination of the "home" position, determination of the actual position of links with respect to reference frame, a kinematic stage, and a dynamic stage. We will use the following notation for describing these stages. N and I denote the sets of indices

$$N = \{i: i \in (1,\ldots,n)\}, \quad I = \{j: j \in (1,\ldots,i)\}$$

where n is the number of the kinematic pairs. ξ_i is a symbol which, depending on the type of kinematic pair i, has the value 0 or 1:

$$\xi_i = \begin{cases} 0 & - \text{ the i-th pair is the revolute one} \\ 1 & - \text{ the i-th pair is the sliding one} \end{cases}$$

Symbol $\bar{\xi}_i = 1-\xi_i$ will also be used, as well as the notation introduced by the preceding definitions.

Stage 1: Determination of "home" position

It is assumed that the active mechanism under consideration may be described by a simple, open kinematic chain Λ_n consisting of a set of kinematic pairs $\{C_{i-1}, C_i\}$, $i \in N$. The kinematic pair $\{C_o, C_1\}$ represents in fact the first link of the mechanism which is connected to support C_o by joint Z_{1o}. It is necessary to determine the position of all links with respect to the reference frame, under the condition that all joint coordinates equal zero, $q_i = 0$, $i \in N$.

Let us assume all kinematic parameters $K_j^o = \left(Q_j^o, R_j^o, E_j^o \right)$ for j $(1, \ldots \ldots, i-1)$ to be known, where the upperscript o means that $q_j = 0$, $j \in N$. The task of determining the "home" position now reduces to determining $K_i^o = \left(Q_i^o, R_i^o, E_i^o \right)$. The sets R_i and E_i for the case of a simple kinematic chain have the following form

$$ R_i = \left\{ \vec{r}_{ii}, \vec{r}_{i,i+1} \right\} \quad \text{and} \quad E_i = \left\{ \vec{e}_i, \vec{e}_{i+1} \right\}, $$

where $\vec{e}_i$ and $\vec{e}_{i+1}$ stand for vectors $\vec{e}_{ii}$ and $\vec{e}_{i,i+1}$.

On the basis of the assumption that K_{i-1}^o is known, it follows that $\vec{e}_i^o$ and $\vec{r}_{i-1,i}^o$ are also known. Further, let us note the set of 3 orthogonal unit vectors in point $Z_{i-1,i}$: $\vec{e}_i^o$, $\vec{a}_i^o$ and $\vec{e}_i^o \times \vec{a}_i^o$, where $\vec{a}_i^o = \text{ort}^{*)} \left(\vec{e}_i^o \times \left(\vec{r}_{i-1,i}^o \times \vec{e}_i^o \right) \right)$, whose components are known with respect to the reference coordinate system (Fig. 2.5). On the other hand, let us note the set of vectors $\tilde{e}_i$, $\tilde{a}_i$ and $\tilde{e}_i \times \tilde{a}_i$, where $\tilde{a}_i = -\text{ort}\left(\tilde{e}_i \times \left(\tilde{r}_{ii} \times \tilde{e}_i \right) \right)$, whose components are known with respect to the local coordinate system Q_i. Under the condition $q_i = 0$, these two sets of vectors coincide. Since $Q_i^o = \left[\vec{q}_{i1}^o \; \vec{q}_{i2}^o \; \vec{q}_{i3}^o \right]$ represents transformation matrix of the i-th link coordinate system into the reference system, it follows that

$$ \vec{e}_i^o = Q_i^o \tilde{e}_i $$

$$ \vec{a}_i^o = Q_i^o \tilde{a}_i \qquad\qquad (2.2.1) $$

$$ \vec{e}_i^o \times \vec{a}_i^o = Q_i^o \left(\tilde{e}_i \times \tilde{a}_i \right) $$

and this completely determines the matrix Q_i^o:

$$ Q_i^o = \left[\vec{e}_i^o \; \vec{a}_i^o \; \vec{e}_i^o \times \vec{a}_i^o \right] \left[\tilde{e}_i \; \tilde{a}_i \; \tilde{e}_i \times \tilde{a}_i \right]^T \qquad (2.2.2) $$

---

*) Ort $(\cdot)$ denotes the unit vector of $(\cdot)$.

Let us note that matrix transposition has been used instead of its inverse, due to the orthogonality of vectors $\tilde{e}_i$, $\tilde{a}_i$ and $\tilde{e}_i \times \tilde{a}_i$.

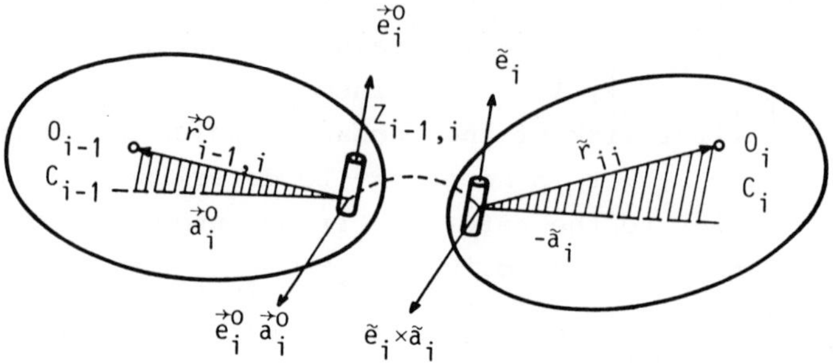

Fig. 2.5. Relative position of link C_i with respect to C_{i-1} with $q_i = 0$

It is now necessary to determine the remaining elements of set K_i^o, i.e. sets R_i^o and E_i^o. Since the transformation matrix Q_i^o has been determined, the following relations evidently hold

$$\vec{r}_{ii}^{\,o} = Q_i^o \tilde{r}_{ii}$$

$$\vec{r}_{i,i+1}^{\,o} = Q_i^o \tilde{r}_{i,i+1} \tag{2.2.3}$$

$$\vec{e}_{i+1}^{\,o} = Q_i^o \tilde{e}_{i+1}$$

Recursiveness required for calculating Q_i^o for $\forall i \in N$, and of all vectors contained in R_i^o and E_i^o, has thus been established.

Stage 2: Mechanism position

This stage consists of determining the elements of the set of kinematic variables $K_i = \left(Q_i,\ R_i,\ E_i \right)$, where the mechanism position is determined by joint coordinates q_i, $i \in N$.

Applying the theorem of finite rotations (Rodrigues' formula) for the case of revolute joints ($\xi_i = 0$) or the theorem of the translation of rigid body ($\xi_i = 1$), one obtains

$$\vec{q}_{ij} = \begin{bmatrix} \vec{e}_i \times \left(\vec{q}_{ij}^{\,o} \times \vec{e}_i \right) \bar{\xi}_i \\[1ex] \left(\vec{e}_i \times \vec{q}_{ij}^{\,o} \right) \bar{\xi}_i \\[1ex] \left(\vec{e}_i \cdot \vec{q}_{ij}^{\,o} \right) \vec{e}_i \bar{\xi}_i + \vec{q}_{ij}^{\,o} \xi_i \end{bmatrix}^T \begin{bmatrix} \cos q_i \\[1ex] \sin q_i \\[1ex] 1 \end{bmatrix}, \quad j = 1, 2, 3. \tag{2.2.4}$$

We have thus determined the matrix $Q_i = \left[\vec{q}_{i1}, \vec{q}_{i2}, \vec{q}_{i3}\right]$ which represents transformation matrix from the i-th link coordinate system S_i into the reference system. Now, we directly obtain

$$R_i = \left\{\vec{r}_{ii}, \vec{r}_{i,i+1}\right\} = \left\{Q_i\tilde{r}_{ii} + q_i\vec{e}_i\xi_i, Q_i\tilde{r}_{i,i+1}\right\}$$

$$E_i = \left\{\vec{e}_i, \vec{e}_{i+1}\right\} = \left\{\vec{e}_i, Q_i\tilde{e}_{i+1}\right\}$$

(2.2.5)

where $\vec{r}_{ij} = \vec{r}_{i-1,j} - \vec{r}_{i-1,i} + \vec{r}_{ii}$, $j \in I$, $i \in N$, evidently holds. Determination of $\tilde{e}_{i+1}$ provides for the recursiveness of calculating the transformation matrix Q_i according to (2.2.4), and thus the elements of sets (2.2.5).

Stage 3: Mechanism kinematics

This stage consists of forming the set of kinematic quantities $\mathcal{L}_i =$ $= \left\{\Omega_i, W_i\right\}$, where

$$\Omega_i = \left\{\vec{\omega}_i\right\}$$

$$W_i = \left\{\vec{\varepsilon}_i, \vec{w}_i\right\}$$

with

$\vec{\omega}_i$ — angular velocity of the i-th link

$\vec{\varepsilon}_i$ — angular acceleration of the i-th link, and

$\vec{w}_i$ — linear acceleration of the i-th link.

These kinematic quantities are functions of the velocities and accelerations of joint coordinates $\dot{q}_i$ and $\ddot{q}_i$ and variables formed in the preceding stage: $K_i = \left(Q_i, R_i, E_i\right)$. All relations considered in this stage are recursive. Applying the basic theorems about rigid body kinematics, one obtains

$$\vec{\omega}_i = \vec{\omega}_{i-1} + \dot{q}_i\vec{e}_i\xi_i$$

$$\vec{\varepsilon}_i = \vec{\varepsilon}_{i-1} + \dot{q}_i\vec{\omega}_{i-1}\times\vec{e}_i\xi_i + \ddot{q}_i\vec{e}_i\xi_i$$

(2.2.6)

$$\vec{w}_i = \vec{w}_{i-1} - \vec{\varepsilon}_{i-1}\times\vec{r}_{i-1,i} - \vec{\omega}_{i-1}\times\left(\vec{\omega}_{i-1}\times\vec{r}_{i-1,i}\right) +$$

$$+ \; \vec{\varepsilon}_i \times \vec{r}_{ii} + \vec{w}_i \times \left(\vec{w}_i \times \vec{r}_{ii} \right) + \left(2\dot{q}_i \vec{w}_{i-1} \times \vec{e}_i + \ddot{q}_i \vec{e}_i \right) \xi_i$$

It follows that angular and linear accelerations are functions of the second derivatives of joint coordinates $\ddot{q}_j$, $j \in I$. However, to form the dynamic model matrices, expressions $\vec{\varepsilon}_i$ and $\vec{w}_i$ should be rearranged so that accelerations $\ddot{q}_j$, $j \in N$, figure explicitly

$$\vec{\varepsilon}_i = \left[\vec{\alpha}_{i1} \; \cdots \; \vec{\alpha}_{ii} \; 0 \; \cdots \; 0 \right] \ddot{q} + \vec{\alpha}_i^{O} \qquad\qquad (2.2.7)$$

$$\vec{w}_i = \left[\vec{\beta}_{i1} \; \cdots \; \vec{\beta}_{ii} \; 0 \; \cdots \; 0 \right] \ddot{q} + \vec{\beta}_i^{O}$$

where $\ddot{q} = \left[\ddot{q}_1 \; \cdots \; \ddot{q}_n \right]$. The set of kinematic quantities W_i thus reduces to

$$W_i = \left\{ \vec{\alpha}_{ij}, \; \vec{\alpha}_i^{O}, \; \vec{\beta}_{ij}, \; \vec{\beta}_i^{O}; \; j \in I \right\}.$$

From (2.2.6) and (2.2.7) we obtain the recursive relations determining the elements of set W_i:

$$\vec{\alpha}_{ij} = \vec{e}_j \bar{\xi}_j, \qquad j \in I, \qquad \vec{\alpha}_i^{O} = \vec{\alpha}_{i-1}^{O\,\cdot} + \dot{q}_i \left(\vec{w}_{i-1} \times \vec{e}_i \right) \xi_i$$

$$\vec{\beta}_{ij} = \vec{\beta}_{i-1,j} + \vec{e}_j \times \vec{R}_{i,i-1} \bar{\xi}_j, \qquad j \in \{1, \ldots, i-1\}$$

$$\vec{\beta}_{ii} = \left(\vec{e}_i \times \vec{r}_{ii} \right) \bar{\xi}_i + \vec{e}_i \xi_i \qquad\qquad (2.2.8)$$

$$\vec{\beta}_i^{O} = \vec{\beta}_{i-1}^{O} + \vec{\alpha}_{i-1}^{O} \times \vec{R}_{i,i-1} + \dot{q}_i \left(\vec{w}_{i-1} \times \vec{e}_i \right) \times \vec{r}_{ii} \bar{\xi}_i \; +$$

$$+ \; 2\dot{q}_i \left(\vec{w}_{i-1} \times \vec{e}_i \right) \xi_i - \vec{\gamma}_{i-1,i} + \vec{\gamma}_{ii}$$

where $\vec{\gamma}_{ij} = \vec{w}_i \times \left(\vec{w}_i \times \vec{r}_{ij} \right)$, $j \in I$. The vector $\vec{R}_{i,i-1} = \vec{r}_{ii} - \vec{r}_{i-1,i}$ represents the distance vector between the $(i-1)$-st and i-th link. All kinematic quantities required for forming the dynamic model have thus been determined.

Stage 4: Mechanism dynamics

This stage includes evaluation of dynamic quantities $\Delta_i = \left\{ F_i, \; M_i \right\}$, $i \in N$, where

$$F_i = \{ \vec{F}_i \}$$

$$M_i = \{ \vec{M}_i \}$$

with

$\vec{F}_i$ - inertial force in the center of mass of the i-th link, and

$\vec{M}_i$ - moment of the inertial force of the i-th link.

The inertial force may be calculated using Newton's law

$$\vec{F}_i = -m_i \vec{w}_i = \left[\vec{a}_{i1} \cdots \vec{a}_{ii} \ 0 \cdots 0 \right] \ddot{q} + \vec{a}_i^o \tag{2.2.9}$$

where m_i - is the mass of the i-th link. Comparing this with (2.2.7), we obtain

$$\vec{a}_{ij} = -m_i \vec{\beta}_{ij}, \qquad j \in I$$
$$\vec{a}_i^o = -m_i \vec{\beta}_i^o. \tag{2.2.10}$$

The moments of inertial forces are determined from Euler's dynamic equations and may be represented in the form

$$\vec{M}_i = \left[\vec{b}_{i1} \cdots \vec{b}_{ii} \ 0 \cdots 0 \right] \ddot{q} + \vec{b}_i^o \tag{2.2.11}$$

where

$$\vec{b}_{ij} = -T_i \vec{\alpha}_{ij}, \qquad j \in I$$
$$\vec{b}_i^o = -T_i \vec{\alpha}_i^o + \vec{\lambda}_i \tag{2.2.12}$$

with

$$T_i = \sum_{\ell=1}^{3} Q_{i\ell} J_{i\ell}$$
$$Q_{i\ell} = \left[q_{i\ell}^1 \vec{q}_{i\ell} \quad q_{i\ell}^2 \vec{q}_{i\ell} \quad q_{i\ell}^3 \vec{q}_{i\ell} \right] \tag{2.2.13}$$

$$\vec{\lambda}_i = Q_i \begin{bmatrix} \left(\vec{\omega}_i \cdot \vec{q}_{i2} \right)\left(\vec{\omega}_i \cdot \vec{q}_{i3} \right)\left(J_{i2} - J_{i3} \right) \\[6pt] \left(\vec{\omega}_i \cdot \vec{q}_{i3} \right)\left(\vec{\omega}_i \cdot \vec{q}_{i1} \right)\left(J_{i3} - J_{i1} \right) \\[6pt] \left(\vec{\omega}_i \cdot \vec{q}_{i1} \right)\left(\vec{\omega}_i \cdot \vec{q}_{i2} \right)\left(J_{i1} - J_{i2} \right) \end{bmatrix}$$

$q_{i\ell}^j$ (j=1,2,3) denoting the j-th component of vector $\vec{q}_{i\ell}$.

Apart from the inertial forces and moments, links are also acted upon by external forces and moments $\vec{G}_i$, $\vec{M}_i^G$, $i \in N$. Thus the total forces and moments may be expressed in the form

$$\vec{F}_i^u = \vec{F}_i + \vec{G}_i$$

$$\vec{M}_i^u = \vec{M}_i + \vec{M}_i^G. \tag{2.2.14}$$

Using the general theorems about rigid body dynamics (theorem of variation of the kinetic moment and theorem of the motion of the center of gravity), differential equations of mechanism motion are obtained. When the i-th kinematic pair is revolute, the first theorem is applied, and when it is a sliding one - the second. Thus, for the revolute kinematic pair ($\xi_i = 0$), one obtains

$$P_i = -\vec{e}_i \cdot \sum_{k=1}^{n} \sum_{j=\max(i,k)}^{n} \left(\vec{b}_{jk} + \vec{r}_{ji} \times \vec{a}_{jk}\right) \ddot{q}_k -$$

$$-\vec{e}_i \cdot \sum_{j=i}^{n} \left(\vec{r}_{ji} \times \left(\vec{a}_j^o + \vec{G}_j\right) + \vec{b}_j^o\right) \tag{2.2.15}$$

where $P_i \in R$ is the driving torque of the i-th actuator. For the sliding kinematic pair ($\xi_i = 1$), one obtains

$$P_i = -\vec{e}_i \cdot \sum_{k=1}^{n} \sum_{j=\max(i,k)}^{n} \vec{a}_{jk} \ddot{q}_k - \vec{e}_i \cdot \sum_{k=1}^{n} \sum_{j=i}^{n} \left(\vec{a}_j^o + \vec{G}_j\right) \tag{2.2.16}$$

Expressions (2.2.15) and (2.2.16) may be written in matrix form

$$P = H(q, \theta)\ddot{q} + h(q, \dot{q}, \theta) \tag{2.2.17}$$

where $P = \begin{bmatrix} P_1 & \cdots & P_n \end{bmatrix}^T$ is the vector of driving torques (forces), $q = \begin{bmatrix} q_1 & \cdots & q_n \end{bmatrix}^T$ is the vector of joint coordinates, $\theta = \begin{bmatrix} \theta_1 & \cdots & \theta_m \end{bmatrix}^T$ is the vector of geometrical and dynamic parameters, $H(q, \theta): R^n \times R^m \to R^{n \times n}$ and $h(q, \dot{q}, \theta): R^n \times R^n \times R^m \to R^n$ are the matrix (vector) of the system which, according to (2.2.15) and (2.2.16), are of the form

$$H(q, \theta) = \begin{bmatrix} H_{ik} \end{bmatrix} = \left[-\vec{e}_i \cdot \sum_{j=\max(i,k)}^{n} \left(\vec{b}_{jk} + \vec{r}_{ji} \times \vec{a}_{jk}\right) \bar{\xi}_i + \vec{a}_{jk} \xi_i \right]$$

$$\tag{2.2.18}$$

$$h(q, \dot{q}, \theta) = \begin{bmatrix} h_i \end{bmatrix} = \left[-\vec{e}_i \cdot \sum_{j=i}^{n} \left(\left(\vec{r}_{ji} \times \left(\vec{a}_j^o + \vec{G}_j\right) + \vec{b}_j^o\right) \bar{\xi}_i + \left(\vec{a}_j^o + \vec{G}_j\right) \xi_i \right) \right]$$

Let us note that the vectors $\vec{e}_i$ and $\vec{r}_{ji}$ contain geometrical parameters, and the vectors $\vec{b}_{jk}$, $\vec{b}_j^o$, $\vec{a}_j^o$ and $\vec{G}_j$ both geometrical and dynamic parameters (the masses and moments of inertia of links). It follows from

(2.2.18) and from all preceding relations that matrix functions $H(q, \theta)$ and $h(q, \dot{q}, \theta)$ are very complex functions of generalized coordinates q and velocities $\dot{q}$ and that they are nonlinear functions of arguments, which results from dynamic interactions of links. The vector θ may be regarded as practically constant.

2.3. Closed-Form Dynamic Model

A model of robot mechanism consisting of a series of rigid bodies interconnected by revolute or sliding joints is under consideration. The same notation as that introduced in the preceding section (configuration, link, kinematic pair and chain etc.) will be used.

Problem statement

Develop the algorithm for forming dynamic robot model in a closed form

$$P = H(q, \theta)\ddot{q} + \dot{q}^T C(q, \theta)\dot{q} + h_G(q, \theta) \tag{2.3.1}$$

where

$\quad H(q, \theta): R^n \times R^m \to R^{n \times n} \quad$ – the inertial matrix of system

$\quad C(q, \theta): R^n \times R^m \to R^{n \times n \times n} \quad$ – "C – matrix" of system, and

$\quad h_G(q, \theta): R^n \times R^m \to R^n \quad$ – gravity vector

with the C – matrix taking into account Coriolis and centrifugal effects, and representing a set of n matrices $\left(C^1(q, \theta), \ldots, C^n(q, \theta)\right)$ such that the following holds

$$\dot{q}^T C(q, \theta)\dot{q} = \begin{bmatrix} \dot{q}^T C^1(q, \theta)\dot{q} \\ \vdots \\ \dot{q}^T C^n(q, \theta)\dot{q} \end{bmatrix} \tag{2.3.2}$$

where $C^i(q, \theta) \in R^{n \times n}$. All matrices of system (2.3.1) should explicitly depend on the following sets of parameters/variables:

$$K_i = \left(Q_i, \ R_i, \ E_i\right), \qquad i \in N$$

$$D_i = \left(m_i, \ J_i\right), \qquad i \in N$$

where $J_i = \left(J_{i1}, \ J_{i2}, \ J_{i3}\right)$ represents the set of the moments of inertia about the principal axes of inertia. The remaining parameters/variables were defined in the preceding section.

Expressions for the matrices of model (2.3.1) should be derived in a closed form.

Problem solution

In developing the algorithm we will use, apart from the notation introduced in the preceding section, the following notation as well:

$$\overline{\left(i_1, \ldots, i_k\right)} = \max\left(i_1, \ldots, i_k\right) - \text{the largest number of } i_1, \ldots, i_k$$

$$n_{ik} = \begin{cases} 1 & \text{for} \quad i > k \\ 0 & \text{for} \quad i < k \end{cases} \quad - \text{ integer function of two indices}$$

$$\overline{n}_{ik} = 1 - n_{ik}$$

$$\overline{F}_{ik} - \text{function defined for } i > k$$

$$\underline{F}_{ik} - \text{function defined for } i < k$$

Let us assume that the position of all links has been determined, i.e. that the elements of set $K_i = \left(Q_i, \ R_i, \ E_i\right)$ are known. We will first derive a nonrecursive form of inertial matrix $H(q, \ \theta)$. For this purpose, let us consider the set of kinematic quantities $\mathcal{L}_i = \left\{\Omega_i, \ W_i\right\}$ defined in stage 3 of the basic Newton-Euler method.

Let us note those elements of set $\mathcal{L}_i$ which refer to derivation of the inertial matrix, i.e. the elements $\vec{\alpha}_{ij}$ and $\vec{\beta}_{ij}$. Summing the recursive expressions for $\vec{\beta}_{\ell j}$, $j \in \{1, \ldots, i-1\}$ according to (2.2.8), one obtains

$$\vec{\beta}_{ij} = \vec{\beta}_{jj} + \vec{e}_j \times \sum_{\ell=j+1}^{i} \overline{R}_{\ell, \ell-1} \vec{\xi}_j \tag{2.3.3}$$

where the summing is carried out with respect to the index ℓ within the limits $j+1$ and i. Evidently, the following holds

$$\sum_{\ell=j+1}^{i} \vec{R}_{\ell,\ell-1} = \vec{R}_{ij}$$

where $\vec{R}_{ij}$ represents the distance vector from the center of mass of the link j to the center of mass of the link i. Taking into account the expression for $\vec{\beta}_{jj}$ from (2.2.8) and substituting into (2.3.3), we obtain

$$\vec{\beta}_{ij} = \left(\vec{e}_j \times \vec{r}_{ij}\right)\bar{\xi}_j + \vec{e}_j \xi_j, \qquad j \in I \qquad (2.3.4)$$

where the relation $\vec{r}_{jj} + \vec{R}_{ij} = \vec{r}_{ij}$ has been used. We will introduce the shortened notation $<\vec{e}, \vec{r}>_{ij}$ for the expression on the right hand side of equation (2.3.4). Thus (2.3.4) becomes

$$\vec{\beta}_{ij} = <\vec{e}, \vec{r}>_{ij} . \qquad (2.3.5)$$

Let us now consider the set of dynamic quantities $\Delta_i = \left\{ F_i, M_i \right\}$ from stage 4 of Newton-Euler method. According to (2.2.10), coefficients $\vec{a}_{ij} \in F_i$ have the following form

$$\vec{a}_{ij} = -m_i <\vec{e}, \vec{r}>_{ij}, \qquad j \in I \qquad (2.3.6)$$

Coefficients $\vec{b}_{ij} \in M_i$, given by the first relation from (2.2.12) and by the first two relations from (2.2.13), may be represented in the form

$$\vec{b}_{ij} = -\sum_{\mu=1}^{3} \left(Q_{i\mu}\vec{e}_j\right) J_{i\mu}\bar{\xi}_j \qquad (2.3.7)$$

Substituting (2.3.6) and (2.3.7) into the expression for the inertial matrix (2.2.18), and using the identities

$$\vec{e}_i \cdot \left(Q_{j\mu}\vec{e}_k\right) \equiv \left(\vec{e}_i \cdot \vec{q}_{j\mu}\right)\left(\vec{e}_k \cdot \vec{q}_{j\mu}\right) \qquad (2.3.8)$$

one obtains

$$H_{ik} = \sum_{j=(i,k)}^{n} \left[m_j <\vec{e}, \vec{r}>_{ji} \cdot <\vec{e}, \vec{r}>_{jk} + \right.$$

$$\left. + \sum_{\mu=1}^{3} \left(\vec{e}_i \cdot \vec{q}_{j\mu}\right)\left(\vec{e}_k \cdot \vec{q}_{j\mu}\right) J_{j\mu}\bar{\xi}_i\bar{\xi}_k \right] \qquad (2.3.9)$$

where H_{ik} is the (i,k)-th element of inertial matrix $H(q, \theta)$.

According to the problem statement, the matrix of centrifugal and Coriolis effects $C(q, \theta)$, defined by (2.3.1) and (2.3.2), should also be determined. Let us assume that the sets of parameters/variables K_i and $\mathcal{D}_i$ are known for $\forall i \in N$, and let us reformulate the elements of sets $\mathcal{L}_i = \{\Omega_i, W_i\}$ from recursive to nonrecursive form. For example, by summing relations (2.2.6) for $\vec{\omega}_j \in \Omega_j$ with respect to the index $j \in I$, one obtains

$$\vec{\omega}_i = \sum_{j=1}^{i} \vec{e}_j \bar{\xi}_j \dot{q}_j \qquad (2.3.10)$$

Let us now consider elements $\vec{\alpha}_i^O \in W_i$ and $\vec{\beta}_i^O \in W_i$. Summing the relations for $\vec{\alpha}_j^O$ over $j \in I$ according to (2.2.8), we obtain

$$\vec{\alpha}_i^O = \sum_{j=1}^{i} \left(\vec{\omega}_{j-1} \times \vec{e}_j\right)\bar{\xi}_j \dot{q}_j . \qquad (2.3.11)$$

Substituting (2.3.10) into (2.3.11), it follows that

$$\vec{\alpha}_i^O = \sum_{k=1}^{i} \sum_{\ell=1}^{k-1} \vec{\varepsilon}_{\ell k} \bar{\xi}_k \bar{\xi}_\ell \dot{q}_k \dot{q}_\ell$$

where $\vec{\varepsilon}_{\ell k} = \vec{e}_\ell \times \vec{e}_k$. Let us denote by $\vec{p}_{\ell k}$ the vector $\vec{\varepsilon}_{\ell k} \bar{\xi}_k \bar{\xi}_\ell \dot{q}_k \dot{q}_\ell$ and consider the following identities

$$\sum_{k=1}^{i} \sum_{\ell=1}^{k-1} \vec{p}_{\ell k} = \sum_{k=1}^{i} \sum_{\ell=1}^{i} \vec{p}_{\ell k} \eta_{k\ell} = \sum_{k=1}^{i} \sum_{\ell=1}^{i} \frac{1}{2}\left(\vec{p}_{\ell k}\eta_{k\ell} + \vec{p}_{k\ell}\eta_{\ell k}\right)$$

Let us note that addends in the last double sum are symmetric with respect to indices k and ℓ. Taking into account $\vec{\varepsilon}_{\ell k} = -\vec{\varepsilon}_{k\ell}$, we obtain

$$\vec{\alpha}_i^O = \sum_{k=1}^{i} \sum_{\ell=1}^{i} \frac{1}{2} \vec{\varepsilon}_{\ell k}\left(\eta_{k\ell} - \eta_{\ell k}\right)\bar{\xi}_k \bar{\xi}_\ell \dot{q}_k \dot{q}_\ell \qquad (2.3.12)$$

It is clear that (2.3.12) may be represented in the form

$$\vec{\alpha}_i^O = \dot{q}^T C_\alpha^i \dot{q} \qquad (2.3.13)$$

where C_α^i is a symmetric n×n matrix:

$$C_\alpha^i = \frac{1}{2}\left[\begin{array}{ccccc|c}
0 & \vec{\varepsilon}_{12}\bar{\xi}_1\bar{\xi}_2 & \vec{\varepsilon}_{13}\bar{\xi}_1\bar{\xi}_3 & \cdots & \vec{\varepsilon}_{1i}\bar{\xi}_1\bar{\xi}_i & \\
\vec{\varepsilon}_{12}\bar{\xi}_1\bar{\xi}_2 & 0 & \vec{\varepsilon}_{23}\bar{\xi}_2\bar{\xi}_3 & \cdots & \vec{\varepsilon}_{2i}\bar{\xi}_2\bar{\xi}_i & \\
\vdots & & & & & 0 \\
\vec{\varepsilon}_{1i}\bar{\xi}_1\bar{\xi}_i & \vec{\varepsilon}_{2i}\bar{\xi}_2\bar{\xi}_i & \vec{\varepsilon}_{3i}\bar{\xi}_3\bar{\xi}_i & \cdots & 0 & \\
\hline
 & & 0 & & & 0
\end{array}\right]$$

We see that the form of expression (2.3.13) is the same as that of (2.3.2) which defines the C - matrix of the system.

Summing relations (2.2.8) for $\vec{\beta}_j^O$ over $j \in I$ and taking into account expression (2.3.12) for $\vec{\alpha}_i^O$, as well as (2.3.10) for $\vec{\omega}_i$, we obtain

$$\vec{\beta}_i^O = \sum_{j=1}^{i} \left\{ \sum_{k=1}^{j-1} \sum_{\ell=1}^{k-1} \vec{\varepsilon}_{\ell k} \times \vec{R}_{j,j-1} \bar{\xi}_k \bar{\xi}_\ell \dot{q}_k \dot{q}_\ell + \right.$$

$$+ \sum_{k=1}^{j-1} \left(\vec{\varepsilon}_{kj} \times \vec{r}_{jj} \bar{\xi}_j + 2 \vec{\varepsilon}_{kj} \xi_j \right) \bar{\xi}_k \dot{q}_j \dot{q}_k +$$

$$+ \sum_{k=1}^{j} \sum_{\ell=1}^{j} \vec{e}_k \times \left(\vec{e}_\ell \times \vec{r}_{jj} \right) \bar{\xi}_k \bar{\xi}_\ell \dot{q}_k \dot{q}_\ell -$$

$$\left. - \sum_{k=1}^{j-1} \sum_{\ell=1}^{j-1} \vec{e}_k \times \left(\vec{e}_\ell \times \vec{r}_{j-1,j} \right) \bar{\xi}_k \bar{\xi}_\ell \dot{q}_k \dot{q}_\ell \right\}$$

Since the reduction of the expression obtained to the simplest possible form is rather complex, this procedure is described in Appendix 2.1. We now give the final result only

$$\vec{\beta}_i^O = \sum_{k=1}^{i} \sum_{\ell=1}^{i} \left(\bar{\xi}_\ell \vec{e}_\ell \times < \vec{e}, \ \vec{r} >_{ik} \eta_{k\ell} + \bar{\xi}_k \vec{e}_k \times < \vec{e}, \ \vec{r} >_{i\ell} \eta_{\ell k} \right) \dot{q}_k \dot{q}_\ell \qquad (2.3.14)$$

This expression may also be presented in the form

$$\vec{\beta}_i^O = \dot{q}^T C_\beta^i \dot{q} \qquad (2.3.15)$$

where C_β^i is a symmetric $n \times n$ matrix. According to (2.3.14), the element (k, ℓ), $k > \ell$ of C_β^i matrix is given by

$$C_\beta^i(k, \ell) = \left(\vec{e}_\ell \times < \vec{e}, \ \vec{r} >_{ik} \right) \bar{\xi}_\ell \eta_{ik}. \qquad (2.3.16)$$

On the basis of (2.3.13) and (2.3.15) we may conclude that the kinematic coefficients $\vec{\alpha}_i^O$ and $\vec{\beta}_i^O$ participate in the formation of C - matrix only. However, these coefficients do not completely determine the C - matrix, but we should also take into account the elements of the set of dynamic parameters/variables $\Delta_i = \left\{ F_i, \ M_i \right\}$. Thus, according to the second relation from (2.2.10), we obtain for $\vec{a}_i^O \in F_i$

$$\vec{a}_i^O = \dot{q}^T C_a^i \dot{q} \qquad (2.3.17)$$

where $C_a^i \in R^{n \times n}$ is a symmetric matrix whose (k, ℓ)-th element $(k > \ell)$ is determined by

$$c_a^i(k, \ell) = -m_i\left(\vec{e}_\ell \times \langle\vec{e}, \vec{r}\rangle_{ik}\right)\bar{\xi}_\ell n_{ik}.$$

(2.3.18)

Let us now consider the set of relations (2.2.11) – (2.2.13) determining $\vec{b}_i^O \in M_i$. Substituting the expression for T_i (2.2.13) and the nonrecursive expression for $\vec{\alpha}_i^O$ (2.3.12) into the expression for $\vec{b}_i^O$ (2.2.12), we obtain

$$\vec{b}_i^O = -\sum_{k=1}^{i} \sum_{\ell=1}^{k-1} \sum_{\ell=1}^{3} Q_{i\mu}\vec{\varepsilon}_{\ell k}J_{i\mu}\bar{\xi}_k\bar{\xi}_\ell\dot{q}_k\dot{q}_\ell + \vec{\lambda}_i$$

(2.3.19)

The expression $Q_{i\mu}\vec{\varepsilon}_{\ell k}$ may be expressed using vectors $\vec{q}_{i\mu}$ and $\vec{\varepsilon}_{\ell k}$

$$Q_{i\mu}\vec{\varepsilon}_{\ell k} = \left(\vec{\varepsilon}_{\ell k}\cdot\vec{q}_{i\mu}\right)\vec{q}_{i\mu}$$

(2.3.20)

where $Q_{i\mu}$ is given by (2.2.13). The vector $\vec{\lambda}_i$ from (2.3.19) is determined by (2.2.13). Substituting the nonrecursive expression for $\vec{\omega}_i$ (2.3.10) and the transformation matrix $Q_i = \left[\vec{q}_{i1}\ \vec{q}_{i2}\ \vec{q}_{i3}\right]$ into the expression for λ_i (2.2.13), one obtains

$$\vec{\lambda}_i = \sum_{k=1}^{i} \sum_{\ell=1}^{i} \vec{\lambda}_{k\ell}^i\dot{q}_k\dot{q}_\ell$$

(2.3.21)

where

$$\vec{\lambda}_{k\ell}^i = \sum_{\mu=1}^{3} \left(\vec{e}_k\cdot\vec{q}_{i[\mu+1]}\right)\left(\vec{e}_\ell\cdot\vec{q}_{i[\mu+2]}\right)\vec{q}_{i\mu}\left(J_{i[\mu+1]}-J_{i[\mu+2]}\right)\bar{\xi}_k\bar{\xi}_\ell$$

(2.3.22)

with $[\mu]$ denoting the function of integer variable $\mu\in\{1,\ldots,5\}$ defined by

$$[\mu] = \mu \bmod 3 = \begin{cases} \mu & 1 \leqslant \mu \leqslant 3 \\ \mu-3 & 4 \leqslant \mu \leqslant 5 \end{cases}$$

By identical transformations (Appendix 2.2) expressions (2.3.21) and (2.3.22) may be reduced to the form

$$\vec{\lambda}_i = \dot{q}^T C_\lambda^i \dot{q}$$

(2.3.23)

where C_λ^i is a symmetric $n \times n$ matrix whose elements for $k > \ell$ are determined by

$$c_\lambda^i(k, \ell) = -\frac{1}{2}\sum_{\mu=1}^{3}\left[\left(\vec{e}_k\cdot\vec{q}_{i\mu}\right)\left(\vec{e}_\ell\times\vec{q}_{i\mu}\right)+\left(\vec{e}_\ell\cdot\vec{q}_{i\mu}\right)\left(\vec{e}_k\times\vec{q}_{i\mu}\right)\right]J_{i\mu}n_{ik}\bar{\xi}_k\bar{\xi}_\ell$$

(2.3.24)

Substituting (2.3.23) and (2.3.24) into (2.3.19), we obtain

$$\vec{b}_i^o = \dot{q}^T C_b^i \dot{q} \tag{2.3.25}$$

where C_b^i is a symmetric n×n matrix whose (k, ℓ)-th element for k>ℓ is given by (Appendix 2.2):

$$C_b^i(k, \ell) = -\frac{1}{2} \sum_{\mu=1}^{3} \left[\left(\vec{\varepsilon}_{\ell k} \cdot \vec{q}_{i\mu} \right) \vec{q}_{i\mu} + \left(\vec{e}_k \cdot \vec{q}_{i\mu} \right) \left(\vec{e}_\ell \times \vec{q}_{i\mu} \right) + \right.$$
$$\left. + \left(\vec{e}_\ell \cdot \vec{q}_{i\mu} \right) \left(\vec{e}_k \times \vec{q}_{i\mu} \right) \right] J_{i\mu} \eta_{ik} \bar{\xi}_k \bar{\xi}_\ell \tag{2.3.26}$$

It was shown in the preceding section that by using the basic theorems of rigid body dynamics the matrices of mathematical model are obtained in the form (2.2.18). Each element h_i of vector $h(q, \dot{q}, \theta)$ may, according to (2.2.18), be represented as

$$h_i = h_i^C + h_i^G \tag{2.3.27}$$

where

$$h_i^C = -\vec{e}_i \cdot \sum_{j=i}^{n} \left(\left(\vec{r}_{ji} \times \vec{a}_j^o + \vec{b}_j^o \right) \bar{\xi}_i + \vec{a}_j^o \xi_i \right) \tag{2.3.28}$$

$$h_i^G = -\vec{e}_i \cdot \sum_{j=i}^{n} \left(\vec{r}_{ji} \times \vec{G}_j \bar{\xi}_i + \vec{G}_j \xi_i \right). \tag{2.3.29}$$

Substituting (2.3.17) and (2.3.25) into (2.3.28), we obtain

$$h_i^C = \dot{q}^T C^i \dot{q} \tag{2.3.30}$$

where $C^i \in R^{n \times n}$ is given by

$$C^i = -\sum_{j=i}^{n} \left\{ \left[\left(\vec{e}_i \times \vec{r}_{ji} \right) \bar{\xi}_i + \vec{e}_i \xi_i \right] \cdot C_a^j + \vec{e}_i \cdot C_b^j \bar{\xi}_i \right\} \tag{2.3.31}$$

with the operation of dot product between the vector and matrix (whose elements are also vectors) reduces to the same operation between the vector and each element of the matrix considered. The matrix C^i evidently represents the i-th component of the C - matrix (See (2.3.2)).

Now, one should substitute the matrices C_a^j (2.3.18) and C_b^j (2.3.26) into the expression on the right-hand side of equation (2.3.31). The element (k, ℓ), k>ℓ, of symmetric matrix C^i is obtained as

$$c_{k\ell}^{i} = \sum_{j=(i,k)}^{n} \left\{ m_j \langle \vec{e}, \vec{r} \rangle_{ji} \cdot \left(\vec{e}_\ell \times \langle \vec{e}, \vec{r} \rangle_{jk} \right) + \right.$$

$$\left. + \frac{1}{2} \sum_{\mu=1}^{3} \left[\left(\vec{e}_i \cdot \vec{q}_{j\mu} \right) \vec{\varepsilon}_{\ell k} + \left(\vec{e}_k \cdot \vec{q}_{j\mu} \right) \vec{\varepsilon}_{i\ell} + \left(\vec{e}_\ell \cdot \vec{q}_{j\mu} \right) \vec{\varepsilon}_{ik} \right] \cdot \vec{q}_{j\mu} J_{j\mu} \bar{\xi}_i \bar{\xi}_k \right\} \bar{\xi}_\ell \qquad (2.3.32)$$

The matrix of centrifugal and Coriolis effects $C(q, \theta)$, defined by (2.3.1) and (2.3.2) has thus been determined.

The gravity vector remains to be determined according to the problem statement. The elements of this vector are given by expression (2.3.29) which is equivalent to

$$h_i^G = -\sum_{j=i}^{n} \left[\left(\left(\vec{e}_i \times \vec{r}_{ji} \right) \bar{\xi}_i + \vec{e}_i \xi_i \right) \cdot \vec{G}_j \right]$$

i.e.,

$$h_i^G = -\sum_{j=i}^{n} \left(\langle \vec{e}, \vec{r} \rangle_{ji} \cdot \vec{G}_j \right) \qquad (2.3.33)$$

which, in fact, represents the solution to the problem, since h_i^G is expressed in terms of the elements of sets $K_i = \left(Q_i, R_i, E_i \right)$ and $\mathcal{D}_i = \left(m_i, J_i \right)$. The vector $\vec{G}_j$ may be expressed as $\vec{G}_j = m_j \vec{g}$, where $\vec{g}$ is a constant vector of gravity acceleration ($|\vec{g}| = 9.81$ m/s^2) given in the reference coordinate system.

The problem stated has thus been solved, i.e. all nonrecursive relations have been derived for forming the matrices of the mathematical model of mechanism whose structure may be described by a simple kinematic chain. It has been assumed that two types of joints exist: sliding and revolute. However, this represents no constraint, since an arbitrary type joint connection may be represented by a series of sliding and revolute joints realized by virtual links of neglectable dimensions and masses.

The algorithm developed is presented in Fig. 2.6. The set of kinematic parameters/variables $K_i = \left(Q_i, R_i, E_i \right)$ is determined using the kinematic stages of Newton-Euler method: the stage of determining "home" and actual mechanism position. Instead, it is also possible to use other methods for constructing the kinematic model of manipulators, such as the method based on Denavit-Hartenberg matrices [18], and the like. The complete algorithm, with all required details, is illustrated in Fig. 2.7.

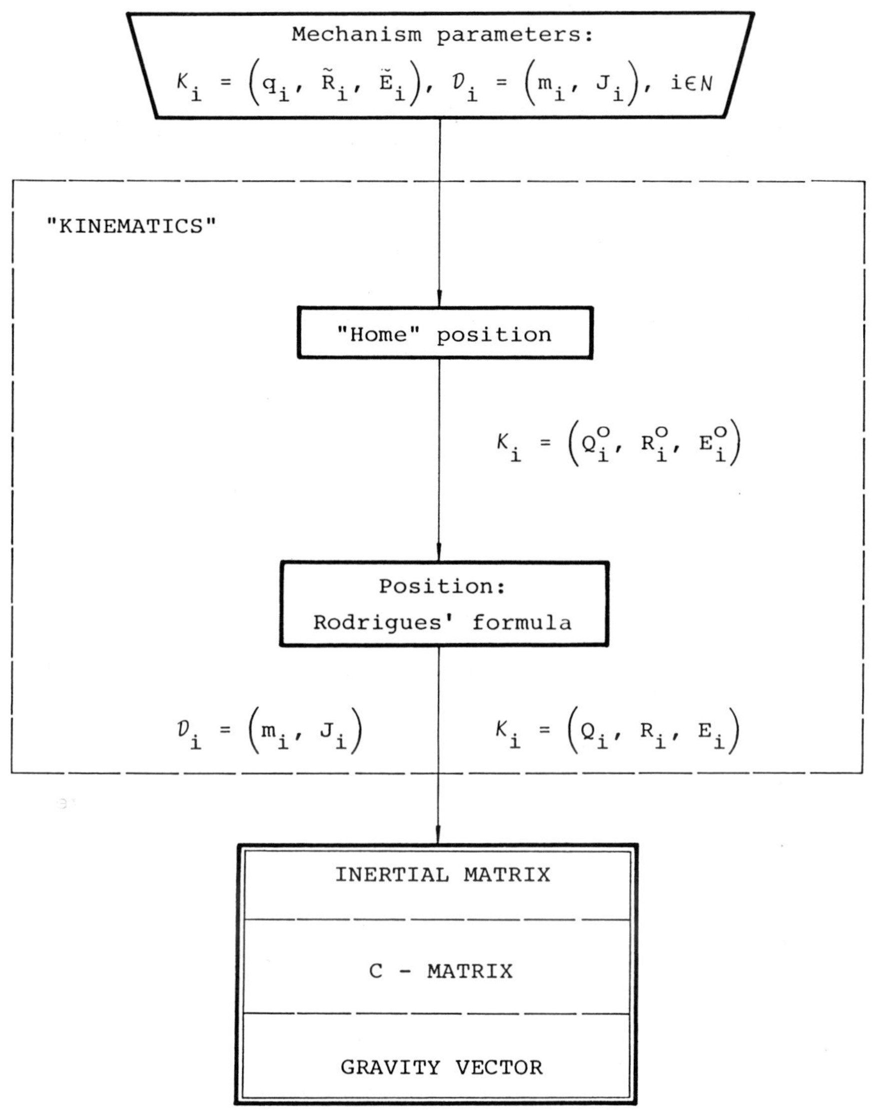

Fig. 2.6. General diagram of the algorithm for forming
the matrices of closed-form dynamic model

INPUT QUANTITIES

n — the number of degrees of freedom

$q = \left(q_1, \ldots, q_n \right)$ — joint coordinates vector

ξ_i — $\begin{cases} 0 - \text{the joint i is the revolute one} \\ 1 - \text{the joint i is the sliding one} \end{cases}$

$\tilde{r}_{ii}$ — the vector of distance from the joint i to the mass center of the link i with respect to the coordinate system of the link i

$\tilde{r}_{i,i+1}$ — the distance vector from the joint i+1 to the mass center of the link i with respect to Q_i

$\tilde{e}_i$ — unit vector of the joint i axis with respect to Q_i

m_i — mass of the link i

$J_{i1}, \; J_{i2}, \; J_{i3}$ — the moments of inertia of the link i with respect to the principal axes of inertia

$\vec{r}_{o1}$ — the distance vector from the first joint to the origin of the reference coordinate system

$\vec{e}_1$ — unit vector of the first joint axis in the reference system

(index $i \in \{1, \ldots, n\}$)

"HOME" POSITION

For $i \in \{1, \ldots, n\}$:

$$\tilde{R}_i = \tilde{e}_i \times \left(\tilde{r}_{ii} \times \tilde{e}_i \right), \qquad \tilde{a}_i = \tilde{R}_i / |\tilde{R}_i|$$

$$\vec{R}_i^{\,o} = \vec{e}_i^{\,o} \times \left(\vec{r}_{i-1,i}^{\,o} \times \vec{e}_i^{\,o} \right), \qquad \vec{a}_i^{\,o} = -\vec{R}_i / |\vec{R}_i|$$

$$\tilde{b}_i = \tilde{e}_i \times \tilde{a}_i, \qquad \vec{b}_i^{\,o} = \vec{e}_i^{\,o} \times \vec{a}_i^{\,o}$$

$$Q_i^O = \begin{bmatrix} \vec{q}_{i1}^{\,O} & \vec{q}_{i2}^{\,O} & \vec{q}_{i3}^{\,O} \end{bmatrix} = \begin{bmatrix} \vec{a}_i^{\,O} & \vec{e}_i^{\,O} & \vec{b}_i^{\,O} \end{bmatrix} \begin{bmatrix} \breve{a}_i & \breve{e}_i & \breve{b}_i \end{bmatrix}^T$$

$$\vec{r}_{ii}^{\,O} = Q_i^O \tilde{r}_{ii}, \qquad \vec{r}_{i,i+1}^{\,O} = Q_i^O \tilde{r}_{i,i+1}$$

$$\vec{e}_{i+1}^{\,O} = Q_i^O \tilde{e}_{i+1}$$

MECHANISM POSITION

For $i \in \{1,\ldots,n\}$:

$$\vec{q}_{ij} = \begin{bmatrix} \vec{e}_i \times \left(\vec{q}_{ij}^{\,O} \times \vec{e}_i \right) \bar{\xi}_i \\[1em] \left(\vec{e}_i \times \vec{q}_{ij}^{\,O} \right) \bar{\xi}_i \\[1em] \left(\vec{e}_i \cdot \vec{q}_{ij}^{\,O} \right) \vec{e}_i \bar{\xi}_i + \vec{q}_{ij}^{\,O} \xi_i \end{bmatrix}^T \begin{bmatrix} \cos q_i \\[1em] \sin q_i \\[1em] 1 \end{bmatrix}, \qquad j=1,2,3$$

$$Q_i = \begin{bmatrix} \vec{q}_{i1} & \vec{q}_{i2} & \vec{q}_{i3} \end{bmatrix}$$

$$\vec{r}_{ii} = Q_i \tilde{r}_{ii} + q_i \vec{e}_i \xi_i, \qquad \vec{r}_{i,i+1} = Q_i \tilde{r}_{i,i+1}$$

$$\vec{e}_{i+1} = Q_i \tilde{e}_{i+1}$$

INERTIAL MATRIX: $H(q,\ \theta)$

$$H_{ik} = \sum_{j=\max(i,k)}^{n} \Big[m_j <\vec{e},\ \vec{r}>_{ji} \cdot <\vec{e},\ \vec{r}>_{jk} +$$

$$+ \sum_{\mu=1}^{3} \left(\vec{e}_i \cdot \vec{q}_{j\mu} \right) \left(\vec{e}_k \cdot \vec{q}_{j\mu} \right) J_{j\mu} \bar{\xi}_i \bar{\xi}_k \Big]$$

$$\Big(H_{ik} = H_{ki} \Big), \qquad i,k \in \{1,\ldots,n\}$$

C - MATRIX: $C(q,\ \theta)$

$$C_{k\ell}^{\,i} = \sum_{\substack{j=\max(i,k)}}^{n} \Big\{ m_j <\vec{e},\ \vec{r}>_{ji} \cdot \left(\vec{e}_\ell \times <\vec{e},\ \vec{r}>_{jk} \right) +$$

$$+ \frac{1}{2} \sum_{\mu=1}^{3} \Big[\left(\vec{e}_i \cdot \vec{q}_{j\mu} \right) \vec{\varepsilon}_{\ell k} + \left(\vec{e}_k \cdot \vec{q}_{j\mu} \right) \vec{\varepsilon}_{i\ell} + \left(\vec{e}_\ell \cdot \vec{q}_{j\mu} \right) \vec{\varepsilon}_{ik} \Big] \cdot \vec{q}_{j\mu} J_{j\mu} \bar{\xi}_i \bar{\xi}_k \Big\} \bar{\xi}_\ell$$

(where $C_{k\ell}^{\,i}$ is for $(k > \ell)$)

$$c^i_{k\ell} = c^i_{\ell k}, \qquad i,k,\ell \in \{1,\dots,n\}$$
$$(k<\ell)$$

GRAVITY VECTOR: $h_G(q, \theta)$

$$h^G_i = -\sum_{j=i}^{n} \left(m_j <\vec{e}, \ \vec{r}>_{ji} \cdot \vec{g} \right), \qquad i \in \{1,\dots,n\}$$

Fig. 2.7. Flowchart of the algorithm for forming the matrices of closed-form dynamic model $\left(<\vec{e}, \ \vec{r}>_{ij} \equiv \vec{e}_j \times \vec{r}_{ij} \vec{\xi}_j + \vec{e}_j \xi_j, \ \vec{\varepsilon}_{ij} \equiv \vec{e}_i \times \vec{e}_j \right)$

2.4. Properties of Dynamic Model Matrices: Symmetry, Positive Definiteness and Antisymmetry

The properties of the matrices $H(q, \theta)$ and $C^i(q, \theta)$ of the dynamic model (2.3.1) - (2.3.2) will be systematized by the following theorems:

Theorem 1.

The inertial matrix $H(q, \theta)$ of dynamic model (2.3.1) is symmetric and positive definite.

Proof:

According to the procedure developed, the element (i, k) of matrix $H(q, \theta)$ is given by (2.3.9). Therefrom, the element (k, i) is

$$H_{ki} = \sum_{j=(k,i)}^{n} \left[m_j <\vec{e}, \ \vec{r}>_{jk} \cdot <\vec{e}, \ \vec{r}>_{ji} + \sum_{\mu=1}^{3} \left(\vec{e}_k \cdot \vec{q}_{j\mu} \right) \left(\vec{e}_i \cdot \vec{q}_{j\mu} \right) J_{j\mu} \xi_k \xi_i \right]$$

Since $\overline{(i, k)} = \overline{(k, i)}$, it follows directly that

$$H_{ki} = H_{ik}, \qquad i \in N, \qquad k \in N$$

which represents the property of symmetry of the inertial matrix.

The second part of the theorem will be proved using the definition of positive definiteness.

Definition

The matrix $H \in R^{n \times n}$ is positive definite if and only if $x^T H x > 0$ holds for each $x \neq 0$, $x \in R^n$.

Let us consider the form of the elements of inertial matrix (2.3.9). Summation in the expression on the right-hand side of (2.3.9) is performed with respect to the index j within the limits $\overline{(i, k)} = \max(i, k)$ and n. Let us extend the sum to the range from 1 to n by introducing integer functions $\bar{n}_{ij}$ and $\bar{n}_{kj}$ defined in the preceding section ($\bar{n}_{ij}=1$ for $j > i$, otherwise 0):

$$H_{ik} = \sum_{j=1}^{n} \left\{ m_j \left(<\vec{e}, \ \vec{r}>_{ji} \bar{n}_{ij} \right) \cdot \left(<\vec{e}, \ \vec{r}>_{jk} \bar{n}_{kj} \right) + \right.$$
$$\left. + \sum_{\mu=1}^{3} \left[\left(\vec{e}_i \cdot \vec{q}_{j\mu} \right) \bar{n}_{ij} \bar{\xi}_i \right] \cdot \left[\left(\vec{e}_k \cdot \vec{q}_{j\mu} \right) \bar{n}_{kj} \bar{\xi}_k \right] J_{j\mu} \right. \tag{2.4.1}$$

Let us introduce the following notation:

$$\vec{p}^{\,0}_{ij} = \sqrt{m_j} \left(<\vec{e}, \ \vec{r}>_{ji} \bar{n}_{ij} \right),$$
$$\vec{p}^{\,\mu}_{ij} = \sqrt{J_{j\mu}} \left(\vec{e}_i \cdot \vec{q}_{j\mu} \right) \bar{n}_{ij} \bar{\xi}_i, \qquad \mu=1,2,3, \tag{2.4.2}$$

and substitute it into (2.4.1). Thus, we obtain

$$H_{ik} = \sum_{j=1}^{n} \sum_{\mu=0}^{3} \vec{p}^{\,\mu}_{ij} \cdot \vec{p}^{\,\mu}_{kj}. \tag{2.4.3}$$

From (2.4.3) it follows that the matrix $H(q, \theta)$ may be represented by

$$H = \sum_{j=1}^{n} \sum_{\mu=0}^{3} \begin{bmatrix} \vec{p}^{\,\mu}_{1j} \cdot \vec{p}^{\,\mu}_{1j} & \cdots & \vec{p}^{\,\mu}_{1j} \cdot \vec{p}^{\,\mu}_{nj} \\ & & \\ & & \\ \vec{p}^{\,\mu}_{nj} \cdot \vec{p}^{\,\mu}_{1j} & \cdots & \vec{p}^{\,\mu}_{nj} \cdot \vec{p}^{\,\mu}_{nj} \end{bmatrix}.$$

or

$$H = \sum_{j=1}^{n} \sum_{\mu=0}^{3} \begin{bmatrix} \vec{p}_{1j}^{\,\mu} \\ \vdots \\ \vec{p}_{nj}^{\,\mu} \end{bmatrix} \cdot \begin{bmatrix} \vec{p}_{1j}^{\,\mu} & \cdots & \vec{p}_{nj}^{\,\mu} \end{bmatrix} \qquad (2.4.4)$$

Introducing the matrix $H_j^{\mu} = \begin{bmatrix} \vec{p}_{1j}^{\,\mu} & \cdots & \vec{p}_{nj}^{\,\mu} \end{bmatrix} \in R^{3 \times n}$, we obtain

$$H = \sum_{j=1}^{n} \sum_{\mu=0}^{3} \left(H_j^{\mu} \right)^T H_j^{\mu} \qquad (2.4.5)$$

Let us denote by $x \in R^n$ an arbitrary vector such that $x \neq 0$, and form the quadratic form

$$x^T H x = x^T \left(\sum_{j=1}^{n} \sum_{\mu=0}^{3} \left(H_j^{\mu} \right)^T H_j^{\mu} \right) x = \sum_{j=1}^{n} \sum_{\mu=0}^{3} x^T \left(H_j^{\mu} \right)^T H_j^{\mu} x =$$

$$= \sum_{j=1}^{n} \sum_{\mu=0}^{3} \left(H_j^{\mu} x \right)^T \left(H_j^{\mu} x \right) \qquad (2.4.6)$$

Since $H_j^{\mu} x \in R^3$ is a vector, and since $x \neq 0$ holds, it follows that

$$\left(H_j^{\mu} x \right)^T \left(H_j^{\mu} x \right) > 0 \qquad (2.4.7)$$

which is a proof that $x^T H x > 0$, i.e. that the inertial matrix is positive definite. Theorem 1 has thus been proved.

Theorem 2.

Submatrices $C^i(q, \theta)$, $i \in N$ of the matrix of Coriolis and centrifugal effects $C(q, \theta)$ of dynamic model (2.3.1) are:

1) symmetric, i.e. $C_{k\ell}^i = C_{\ell k}^i$, $k, \ell \in N$, and

2) antisymmetric with respect to indices i and k, i.e. it holds

$$C_{k\ell}^i = -C_{i\ell}^k, \qquad \text{for} \quad k, i > \ell, \qquad (2.4.8)$$

with $C_{k\ell}^i = 0$ applying for $i = k > \ell$.

Proof:

According to the derivation from the preceding section matrices $C_a^j \in R^{n \times n}$ and $C_b^j \in R^{n \times n}$ are symmetric, so the matrix $C^i \in R^{n \times n}$ obtained from (2.3.31) is also a symmetric one.

To prove property 2), let us start from expression (2.3.32) and form $c_{i\ell}^k$:

$$c_{i\ell}^{k} = \sum_{\substack{(i>\ell) \\ j=\overline{(k,i)}}}^{n} \left\{ m_j <\vec{e},\ \vec{r}>_{jk} \cdot \left(\vec{e}_\ell \times <\vec{e},\ \vec{r}>_{ji} \right) + \right.$$
$$\left. + \frac{1}{2} \sum_{\mu=1}^{3} \left[\left(\vec{e}_k \cdot \vec{q}_{j\mu} \right) \vec{\varepsilon}_{\ell i} + \left(\vec{e}_i \cdot \vec{q}_{j\mu} \right) \vec{\varepsilon}_{k\ell} + \left(\vec{e}_\ell \cdot \vec{q}_{j\mu} \right) \vec{\varepsilon}_{ki} \right] \cdot \vec{q}_{j\mu} J_{j\mu} \bar{\xi}_i \bar{\xi}_k \right\} \bar{\xi}_\ell \qquad (2.4.9)$$

Since the following holds

$$<\vec{e},\ \vec{r}>_{jk} \cdot \left(\vec{e}_\ell \times <\vec{e},\ \vec{r}>_{ji} \right) = -<\vec{e},\ \vec{r}>_{ji} \cdot \left(\vec{e}_\ell \times <\vec{e},\ \vec{r}>_{jk} \right),$$

$$\vec{\varepsilon}_{ij} = -\vec{\varepsilon}_{ji}, \qquad i,j \in N, \quad \text{and} \cdot$$

$$\overline{(k,\ i)} = \overline{(i,\ k)},$$

by substituting into (2.4.9) and comparing with (2.3.32), the relation of antisymmetry (2.4.8) is directly obtained. If i=k, it is directly recognizable that the expression under sum in (2.4.9) equals 0, which is a proof of Theorem 2.

These theorems are important since they describe the main properties of dynamic model matrices and since their application considerably reduces the number of operations required for the construction of non-recursive model according to the flowchart in Fig. 2.7. To illustrate this, let us say that, by applying the property of symmetry, the number of elements of the inertial matrix that should be computed reduces from n^2 to

$$n_H = \frac{n(n+1)}{2}.$$

It is easy to prove that, by applying the symmetry and antisymmetry properties of C - matrices, the number of elements of these matrices reduces from n^3 to

$$n_C = \frac{n(n^2-1)}{3}.$$

For example, for n=6 this number reduces from 216 to 70, i.e. the number of elements is reduced to 32.4% of the original number.

Finally, it should be stated that the properties proved in this section are rather difficult to prove using classical recursive algorithms.

2.5. Closed-Form Linearized Model

The dynamic model (2.3.1) of n-degrees-of-freedom robot manipulator (2.3.1) represents a system of n complex, nonlinear differential equations with respect to the vector of joint coordinates q. In addition, the system is nonlinear with respect to argument $\dot{q}$, as a result of centrifugal and Coriolis forces. That's why the problem of computer--aided design of the linearized robot model is to be introduced.

The total differential of the system (2.3.1) may be represented in the form

$$\delta P = H(q,\ \theta)\,\delta\ddot{q} + 2[\dot{q}^T C(q,\ \theta)]\,\delta\dot{q} + D(q,\ \dot{q},\ \ddot{q},\ \theta)\,\delta q \qquad (2.5.1)$$

where

$$D = \frac{\partial H(q,\ \theta)}{\partial q}\,\ddot{q} + \dot{q}^T\,\frac{\partial C(q,\ \theta)}{\partial q}\,\dot{q} + \frac{\partial h_G(q,\ \theta)}{\partial q} \qquad (2.5.2)$$

The matrix $[\dot{q}^T C(q,\theta)] \in R^{n\times n}$ has the following structure

$$[\dot{q}^T C(q,\ \theta)] = \begin{bmatrix} \dot{q}^T C^1(q,\ \theta) \\ \vdots \\ \dot{q}^T C^n(q,\ \theta) \end{bmatrix} \qquad (2.5.3)$$

The problem of linearized model construction may be formulated as follows:

Problem statement

Develop a procedure for forming the matrix $D \in R^{n\times n}$ of linearized model (2.5.1). The procedure should be based on the algorithm for constructing closed-form dynamic model presented in Para. 2.3. In forming the matrices of partial derivatives

$$\frac{\partial H(q,\ \theta)}{\partial q}\ ,\qquad \frac{\partial C(q,\ \theta)}{\partial q}\ ,\qquad \frac{\partial h_G(q,\ \theta)}{\partial q} \qquad (2.5.4)$$

one should use such operations which do not accumulate errors, e.g.
numerical differentation, and the like.

Problem solution

The same notation as that given in 2.3 will be used to develop the algorithm

$$N = \{1,\ldots,n\}, \qquad I = \{1,\ldots,i\},$$

$$\eta_{ij} = \begin{cases} 1 & \text{for} \quad i>j \\ 0 & \text{for} \quad i<j \end{cases}, \qquad i\in N, \ j\in N,$$

$$\bar{\eta}_{ij} = 1 - \eta_{ij}.$$

The problem will be solved using some properties of open kinematic
chains, which will be summarized by three theorems:

Theorem 1.

The partial deviative of a noncomposite vector $\vec{a}_i$ (a noncomposite vector $\vec{a}_i$ is a vector which is constant with respect to the local coordinate system of the link i), with respect to joint coordinate q_s, $s\in N$,
is given by

$$\frac{\partial \vec{a}_i}{\partial q_s} = \left(\vec{e}_s \times \vec{a}_i\right)\bar{\xi}_s\bar{\eta}_{si} \tag{2.5.5}$$

where $\vec{e}_s$ is the unit vector of joint axis of the kinematic pair s.

Proof:

When the kinematic pair s is revolute ($\xi_s=0$) and s<i ($\bar{\eta}_{si}=1$), the problem of determining the partial derivative

$$\frac{\Delta \vec{a}_i}{\Delta q_s} = \lim_{\Delta q_s \to 0} \frac{\Delta \vec{a}_i}{\Delta q_s}\bigg|_{\substack{\Delta q_j=0 \\ (j\neq s)}}$$

reduces to the well-known theorem of elementary rotations from theoretical mechanics. According to this theorem

$$\lim_{\Delta q_s \to 0} \frac{\Delta \vec{a}_i}{\Delta q_s} = \vec{e}_s \times \vec{a}_i$$

if rotation occurs about the axis $\vec{e}_s$ only and if $\vec{a}_i$ is a noncomposite vector.

If $s > i$, in rotating about the axis s the vector $\vec{a}_i$ does not change its position, so the partial derivative of $\vec{a}_i$ with respect to q_s equals zero. In addition, for the case of sliding pair ($\xi_\ell = 1$), the vector $\vec{a}_i$ does not change, i.e. $\partial \vec{a}_i / \partial q_s = 0$, and this is the proof of the theorem.

Theorem 2.

The partial derivative of vector $\vec{r}_{ij}$, $i \in N$, $i > j$, connecting the joint of the j-th kinematic pair and the center of gravity of the i-th link, with respect to the joint coordinate q_s, $s \in N$, is given by the expression

$$\frac{\partial \vec{r}_{ij}}{\partial q_s} = \left(\vec{e}_s \times \vec{r}_{i,\overline{(j,s)}} \xi_s + \vec{e}_s \xi_s \eta_{sj} \right) \bar{\eta}_{si} \qquad (2.5.6)$$

where $\vec{e}_s$ is the unit vector of the s-th joint axis.

Proof:

Let us consider three cases.

1) $s < j$. In this case $\vec{r}_{ij}$ behaves as a noncomposite vector. Thus Theorem 1 is directly applicable. Since the value of η_{sj} is equal to zero for $s < j$, the proof is completed.

2) $j < s < i$. In this case we may decompose the vector into two components

$$\vec{r}_{ij} = \vec{r}_{is} + \vec{R}_{sj}$$

as shown in Fig. 2.8. Let us assume that the kinematic pair $\left\{ C_{s-1}, C_s \right\}$ is revolute, and that q_s changes by Δq_s. $\vec{R}_{sj}$ then remains unchanged, and the vector $\vec{r}_{is}$ rotates about the axis $\vec{e}_s$. Thus, Theorem 1 can also be applied to it. Since $\overline{(j, s)} = s$ for $s > j$, expression (2.5.6) reduces to the corresponding expression from Theorem 1.

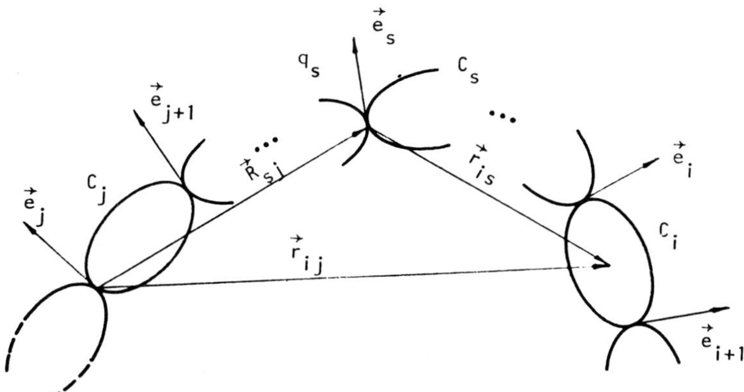

Fig. 2.8. Some characteristic vectors of kinematic chain

If the kinematic pair $\left(C_{s-1},\ C_s\right)$ is sliding, upon a change of q_s by Δq_s the vector $\vec{r}_{is}$ remains unchanged and the vector $\vec{R}_{si}$ changes by $\vec{e}_s \Delta q_s$. Accordingly, the partial derivative of vector $\vec{r}_{ij}$ with respect to q_s reduces to the vector $\vec{e}_s$, and this is given by expression (2.5.6).

3) $s>i$. In this case, the vector $\vec{r}_{ij}$ remains unchanged upon a change of q_s, so the partial derivative of $\vec{r}_{ij}$ is 0, and this is just what is obtained from (2.5.6).

This completes the proof of the theorem.

Theorem 3.

The partial derivative of a cross or dot product of two vectors (which may be composite) $\vec{a}_i$ and $\vec{b}_i$, $i \in N$, is given by

$$\frac{\partial}{\partial q_s}\left(\vec{a}_i \times \vec{b}_i\right) = \frac{\partial \vec{a}_i}{\partial q_s} \times \vec{b}_i + \vec{a}_i \times \frac{\partial \vec{b}_i}{\partial q_s}, \qquad s \in N. \tag{2.5.7}$$

Since this is an elementary theorem of vector analysis, it will not be proved here.

In order to form the linearized model (2.5.1) - (2.5.2), let us consider the nonrecursive forms of dynamic model matrices obtained in Para. 2.3. (Fig. 2.7). The expressions for inertial matrix, C - matrix and gravity vector contain the following elements

$$<\vec{e},\ \vec{r}>_{ij} = \vec{e}_j \times \vec{r}_{ij} \bar{\xi}_j + \vec{e}_j \xi_j \qquad (i>j) \tag{2.5.8}$$

$$\vec{\varepsilon}_{ij} = \vec{e}_i \times \vec{e}_j, \qquad (j>i) \tag{2.5.9}$$

and the scalar

$$P_{ij\mu} = \vec{e}_i \cdot \vec{q}_{j\mu}, \qquad (j>i) \tag{2.5.10}$$

Let us first apply Theorem 3, and then Theorems 1 and 2 to expression (2.5.8):

$$\langle \vec{e}, \vec{r} \rangle_{ij}^{(s)} = \left(\vec{e}_j^{(s)} \times \vec{r}_{ij} + \vec{e}_j \times \vec{r}_{ij}^{(s)} \right) \bar{\xi}_j + \vec{e}_j^{(s)} \xi_j =$$

$$= \left(\left(\vec{\varepsilon}_{sj} \times \vec{r}_{ij} + \vec{\varepsilon}_{sj} \right) \bar{\eta}_{sj} \bar{\xi}_s + \left(\vec{e}_j \times \left(\vec{e}_s \times \vec{r}_{i, \overline{(j,s)}} \right) \bar{\xi}_s + \right.$$

$$+ \left. \vec{\varepsilon}_{js} \xi_s \eta_{sj} \right) \bar{\eta}_{si} \right) \bar{\xi}_j \tag{2.5.11}$$

where $(\cdot)^{(s)}$ denotes the partial derivative of variable $(\cdot)$ with respect to q_s. The following expression is equivalent to expression (2.5.11)

$$\langle \vec{e}, \vec{r} \rangle_{ij}^{(s)} \atop (i>j) = \begin{cases} \vec{e}_s \times \left(\vec{e}_j \times \vec{r}_{ij} + \vec{e}_j \right) \bar{\xi}_s \bar{\xi}_j & s<j \\[2ex] \vec{e}_j \times \langle \vec{e}, \vec{r} \rangle_{is} \bar{\xi}_j & j<s<i \\[2ex] 0 & s>i \end{cases} \tag{2.5.12}$$

Applying Theorem 3 and Theorem 1 to expression (2.5.9), we obtain

$$\vec{\varepsilon}_{ij}^{(s)} = \vec{e}_i^{(s)} \times \vec{e}_j + \vec{e}_i \times \vec{e}_j^{(s)} = \left(\vec{\varepsilon}_{si} \times \vec{e}_j \bar{\eta}_{si} + \vec{e}_i \times \vec{\varepsilon}_{sj} \bar{\eta}_{sj} \right) \bar{\xi}_s \tag{2.5.13}$$

or,

$$\vec{\varepsilon}_{ij}^{(s)} \atop (j>i) = \begin{cases} \vec{e}_s \times \vec{\varepsilon}_{ij} \bar{\xi}_s & s<i \\[2ex] \vec{e}_i \times \vec{\varepsilon}_{sj} \bar{\xi}_s & i<s<j \\[2ex] 0 & s>j \end{cases} \tag{2.5.14}$$

In the same way expression (2.5.10) yields

$$P_{ij\mu}^{(s)} = \vec{e}_i^{(s)} \cdot \vec{q}_{j\mu} + \vec{e}_i \cdot \vec{q}_{j\mu}^{(s)} = \vec{q}_{j\mu} \cdot \vec{\varepsilon}_{si} \left(\bar{\eta}_{si} - \bar{\eta}_{sj} \right) \bar{\xi}_s \tag{2.5.15}$$

Let us now obtain the partial derivatives of inertial matrix elements by applying Theorem 3:

$$\frac{\partial H_{ik}}{\partial q_s} = \sum_{j=\max(i,k)}^{n} \left[m_j \left(<\vec{e}, \ \vec{r}>_{ji}^{(s)} \cdot <\vec{e}, \ \vec{r}>_{jk} + <\vec{e}, \ \vec{r}>_{ji} \cdot <\vec{e}, \ \vec{r}>_{jk}^{(s)} \right) + \right.$$

$$\left. + \sum_{\mu=1}^{3} \left(p_{ij\mu}^{(s)} p_{kj\mu} + p_{ij\mu} p_{kj\mu}^{(s)} \right) J_{j\mu} \bar{\xi}_i \bar{\xi}_k \right] \qquad (2.5.16)$$

Expressions for the partial derivatives of the elements of C - matrix and of gravity vector are obtained in a similar way. The problem stated has thus been solved.

Finally, it should be stated that the construction of linearized model requires not only the determination of the matrices of partial derivatives (2.5.4) but also of the matrices of dynamic model $H(q, \theta)$ and $C(q, \theta)$, as may be seen from (2.5.1).

A general flow-chart for the construction of nonlinear and linearized model is given in Fig. 2.9. All expressions are given in Fig. 2.10.

2.6. Closed-Form Sensitivity Model

Let us consider the dependence of the matrices of dynamic model (2.3.1) on the vector of parameters θ. The vector of parameters consists of the kinematic parameters

$$\tilde{R}_i = \left\{ \tilde{r}_{ii}, \ \tilde{r}_{i,i+1} \right\} \subset K_i,$$
$$\qquad (2.6.1)$$
$$\tilde{E}_i = \left\{ \tilde{e}_i \right\} \subset K_i,$$

which are given with respect to local coordinate systems Q_i, $i \in N$, and of the dynamic parameters

$$\mathcal{D}_i = \left\{ m_i, \ J_{i1}, \ J_{i2}, \ J_{i3} \right\}, \qquad i \in N. \qquad (2.6.2)$$

The influence of variations in kinematic parameters upon the matrices of dynamic model (see the diagram in Fig. 2.7) is nonlinear. However, these parameters may often be regarded as constant, since they represent in fact the geometrical parameters of the mechanism. In cases where they do vary, one should take into account the fact that their influence on the elements of inertial and C - matrix is of the type of quadratic function and on the elements of gravity vector - linear.

In this section we will consider the case involving finite variations of dynamic parameters (2.6.2). Although the variation in dynamic parameters occurs most frequently with the n-th link, because of the influence of workpiece parameters, let us denote by s the number of the link with which the variation occurs:

$$\Delta\theta_s = \left\{\Delta m_s,\ \Delta J_{s1},\ \Delta J_{s2},\ \Delta J_{s3}\right\}. \tag{2.6.3}$$

It is now possible to state the following problem:

Problem statement

Let the robot mechanism model be described by (2.3.1). It is necessary to determine the variation ΔP in the vector of driving torques (forces) upon a variation $\Delta\theta_s$ in the vector of dynamic parameters of link s (2.6.3), assuming that the vectors q, $\dot{q}$ and $\ddot{q}$ do not change.

Problem solution

Let us consider expression (2.3.9) for the element H_{ik} of inertial matrix $H(q,\ \theta)$. Let us assume that the dynamic parameters of link s have changed by Δm_s and $\Delta J_{s\mu}$ ($\mu = 1,2,3$). We obtain from (2.3.9)

$$\Delta H_{ik} = \left(\langle \vec{e},\ \vec{r}\rangle_{si}\cdot\langle \vec{e},\ \vec{r}\rangle_{sk}\Delta m_s + \right.$$

$$\left. + \sum_{\mu=1}^{3}\left(\vec{e}_i\cdot\vec{q}_{s\mu}\right)\left(\vec{e}_k\cdot\vec{q}_{s\mu}\right)\Delta J_{s\mu}\bar{\xi}_i\bar{\xi}_k\right)\bar{\eta}_{\overline{(i,k)},s} \tag{2.6.4}$$

Let us introduce the following notation

$$H_s^O(i,k) = \langle \vec{e},\ \vec{r}\rangle_{si}\cdot\langle \vec{e},\ \vec{r}\rangle_{sk}\bar{\eta}_{\overline{(i,k)},s} \tag{2.6.5}$$

$$H_s^\mu(i,k) = \left(\vec{e}_i\cdot\vec{q}_{s\mu}\right)\left(\vec{e}_k\cdot\vec{q}_{s\mu}\right)\bar{\xi}_i\bar{\xi}_k\bar{\eta}_{\overline{(i,k)},s} \tag{2.6.6}$$

for $\mu = 1,2,3$. Substituting into (2.6.4), we obtain

$$\Delta H_{ik} = \sum_{\mu=0}^{n_d} H_s^\mu(i,k)\Delta\theta_s(\mu+1) \tag{2.6.7}$$

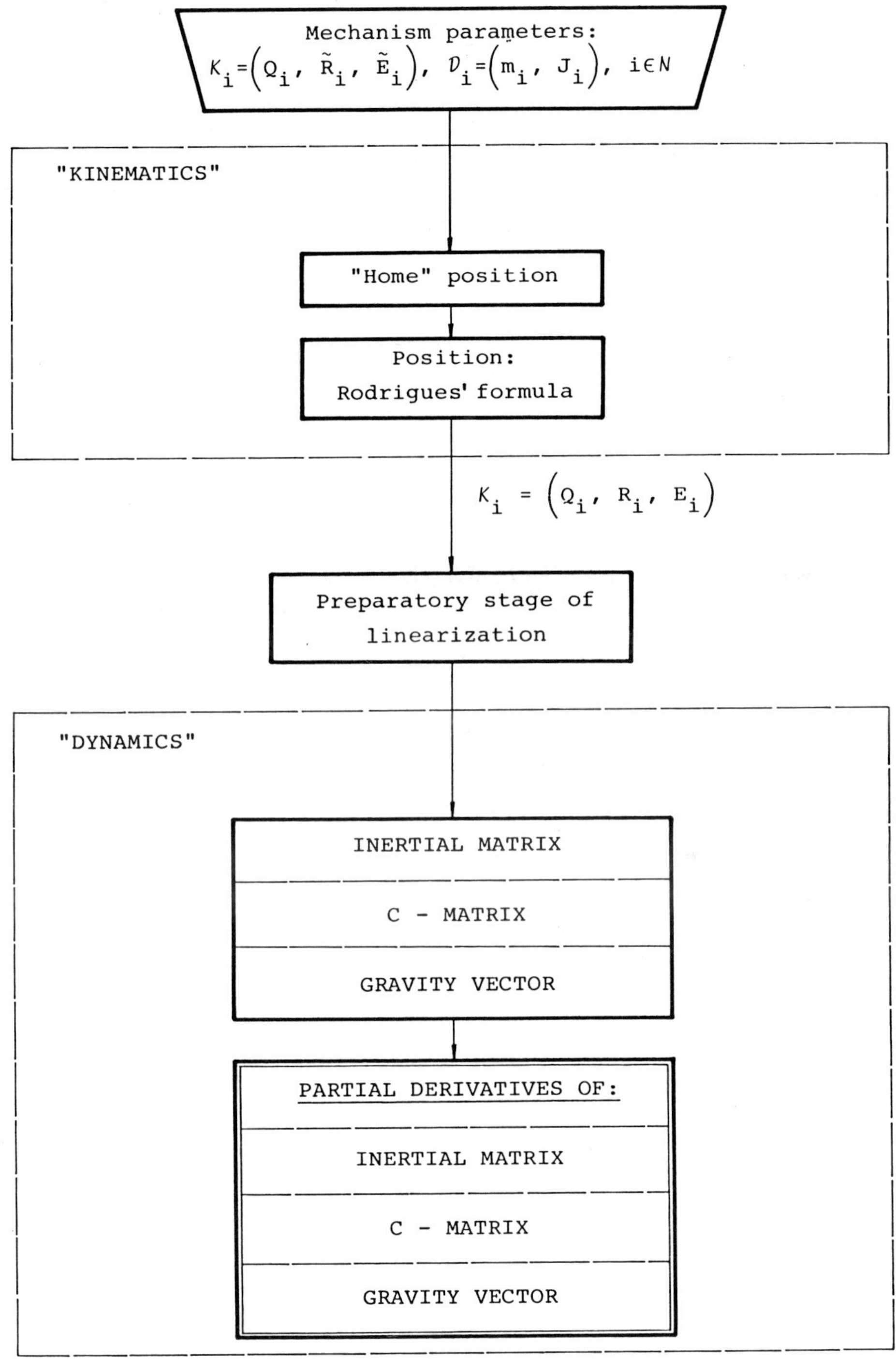

Fig. 2.9. General flow-chart for constructing nonlinear
and linearized model

92

"Preparatory stage of linearization"

$$\langle \vec{e},\ \vec{r} \rangle^{(s)}_{ij} = \left(\left(\vec{\varepsilon}_{sj} \times \vec{r}_{ij} + \vec{\varepsilon}_{sj} \right) \bar{\eta}_{sj} \bar{\xi}_s + \left(\vec{e}_j \times \left(\vec{e}_s \times \vec{r}_{i,\overline{(j,s)}} \right) \bar{\xi}_s + \vec{\varepsilon}_{js} \xi_s \eta_{sj} \right) \bar{\eta}_{si} \right) \bar{\xi}_j$$

$$\vec{\varepsilon}^{(s)}_{ji} = \left(\vec{\varepsilon}_{si} \times \vec{e}_j \bar{\eta}_{si} + \vec{e}_i \times \vec{\varepsilon}_{sj} \bar{\eta}_{sj} \right) \bar{\xi}_s$$

$$p^{(s)}_{ij\mu} = \vec{q}_{j\mu} \cdot \vec{\varepsilon}_{si} \left(\bar{\eta}_{si} - \bar{\eta}_{sj} \right) \bar{\xi}_s$$

"Partial derivatives of inertial matrix,
C - matrix and gravity vector"

$$H^{(s)}_{ik} = \sum_{j=\max(i,k)}^{n} \left[m_j \left(\langle \vec{e},\ \vec{r} \rangle \right)^{(s)}_{ji} \cdot \langle \vec{e},\ \vec{r} \rangle_{jk} + \langle \vec{e},\ \vec{r} \rangle_{ji} \cdot \langle \vec{e},\ \vec{r} \rangle^{(s)}_{jk} \right.$$

$$\left. + \sum_{\mu=1}^{3} \left(p^{(s)}_{ij\mu} p_{kj\mu} + p_{ij\mu} p^{(s)}_{kj\mu} \right) J_{j\mu} \bar{\xi}_i \bar{\xi}_k \right]$$

$$\left(C^i_{k\ell} \right)^{(s)}_{(k>\ell)} = \sum_{j=\max(i,k)}^{n} \left\{ m_j \left[\langle \vec{e},\ \vec{r} \rangle^{(s)}_{ji} \cdot \left(\vec{e}_\ell \times \langle \vec{e},\ \vec{r} \rangle_{jk} \right) \right. \right.$$

$$+ \langle \vec{e},\ \vec{r} \rangle_{ji} \cdot \left(\vec{\varepsilon}_{s\ell} \times \langle \vec{e},\ \vec{r} \rangle_{jk} \right) \xi_s \bar{\eta}_{s\ell} + \langle \vec{e},\ \vec{r} \rangle_{ji} \cdot \left(\vec{e}_\ell \times \langle \vec{e},\ \vec{r} \rangle^{(s)}_{jk} \right) \right]$$

$$+ \frac{1}{2} \sum_{\mu=1}^{3} \left[\left(p^{(s)}_{ij\mu} \vec{\varepsilon}_{\ell k} + p^{(s)}_{kj\mu} \vec{\varepsilon}_{i\ell} + p^{(s)}_{\ell j\mu} \vec{\varepsilon}_{ik} + p_{ij\mu} \vec{\varepsilon}^{(s)}_{\ell k} \right. \right.$$

$$\left. + p_{kj\mu} \vec{\varepsilon}^{(s)}_{i\ell} + p_{\ell j\mu} \vec{\varepsilon}^{(s)}_{ik} \right) \cdot \vec{q}_{j\mu} + \left(p_{ij\mu} \vec{\varepsilon}_{\ell k} + p_{kj\mu} \vec{\varepsilon}_{i\ell} \right.$$

$$\left. \left. + p_{\ell j\mu} \vec{\varepsilon}_{ik} \right) \cdot \left(\vec{e}_s \times \vec{q}_{j\mu} \right) \xi_s \bar{\eta}_{sj} \right] J_{j\mu} \bar{\xi}_i \bar{\xi}_k \right\} \bar{\xi}_\ell$$

$$\left(h^G_i \right)^{(s)} = - \sum_{j=i}^{n} \left(m_j \langle \vec{e},\ \vec{r} \rangle^{(s)}_{ji} \cdot \vec{g} \right)$$

Fig. 2.10. Algorithm for linearized model construction

where $\Delta\theta_s(1) = \Delta m_s$, $\Delta\theta_s(\mu+1) = \Delta J_{s\mu}$, $(\mu = 1,2,3)$, $n_d = 3$, or in matrix form:

$$\Delta H(q, \theta) = \sum_{\mu=0}^{n_d} H_s^\mu(q) \Delta\theta_s(\mu+1), \tag{2.6.8}$$

where $H_s^\mu(q): R^n \to R^{n \times n}$.

The C - matrix (2.3.32) will be treated in a similar way. Under the same assumption on parameter variation, one obtains

$$\Delta C_{k\ell}^i = \sum_{\mu=0}^{n_d} C_s^{i\mu}(k,\ell) \Delta\theta_s(\mu+1), \tag{2.6.9}$$

where

$$C_s^{io}(k,\ell) = \langle \vec{e}, \vec{r} \rangle_{si} \cdot \left(\vec{e}_\ell \times \langle \vec{e}, \vec{r} \rangle_{sk} \right) \bar{\xi}_\ell \bar{\eta}_{\overline{(i,k)},s} \tag{2.6.10}$$

$$C_s^{i\mu}(k,\ell) = \frac{1}{2} \left[\left(\vec{e}_i \cdot \vec{q}_{s\mu} \right) \vec{\varepsilon}_{\ell k} + \left(\vec{e}_k \cdot \vec{q}_{s\mu} \right) \vec{\varepsilon}_{i\ell} + \right.$$

$$\left. + \left(\vec{e}_\ell \cdot \vec{q}_{s\mu} \right) \vec{\varepsilon}_{ik} \right] \cdot \vec{q}_{s\mu} \bar{\xi}_i \bar{\xi}_k \bar{\xi}_\ell \bar{\eta}_{\overline{(i,k)},s} \tag{2.6.11}$$

for $k > \ell$. Expression (2.6.9) may also be expressed in matrix form

$$\Delta C^i(q, \theta) = \sum_{\mu=0}^{n_d} C_s^{i\mu}(q) \Delta\theta_s(\mu+1), \tag{2.6.12}$$

where $C_s^{i\mu}(q): R^n \to R^{n \times n}$.

Let us now consider the parametric dependence of gravity vector $h_G(q,\theta)$. Taking into account that in (2.3.33) the vector $\vec{G}_j$ is given by $m_j \vec{g}$, we obtain

$$\Delta h_i^G = \langle \vec{e}, \vec{r} \rangle_{si} \cdot \vec{g} \bar{\eta}_{is} \Delta m_s \tag{2.6.13}$$

Introducing the notation

$$h_s^G(i) = \langle \vec{e}, \vec{r} \rangle_{si} \cdot \vec{g} \bar{\eta}_{is} \tag{2.6.14}$$

and substituting (2.6.14) into (2.6.13), we obtain

$$\Delta h_i^G = h_s^G(i) \Delta m_s . \tag{2.6.15}$$

In vector form, this relation becomes

$$\Delta h^G(q,\ \theta)\ =\ \sum_{\mu=0}^{n_d}\ h_s^{G\mu}(q)\,\Delta\theta_s\,(\mu+1) \qquad (2.6.16)$$

where $h_s^{G\mu}(q):\ R^n \to R^n$, and

$$h_s^{Go}\ =\ \left[\,h_s^G(1)\ \cdots\ h_s^G(n)\,\right]^T, \qquad (2.6.17)$$

$$h_s^{G\mu}\ =\ [0\ \cdots\ 0]^T \epsilon R^n, \qquad \mu\ =\ 1,2,3. \qquad (2.6.18)$$

According to the problem statement, it is necessary to determine ΔP upon parameter variation $\Delta\theta_s$, with q, $\dot{q}$ and $\ddot{q}$ given. From (2.6.8), (2.6.12) and (2.6.16), it follows that the influence of $\Delta\theta_s$ on model matrices is linear. As a result of linearity from (2.3.1), one obtains

$$\Delta P\ =\ \Delta H(q,\ \theta)\ddot{q}\ +\ \dot{q}^T\Delta C(q,\ \theta)\dot{q}\ +\ \Delta h^G(q,\ \theta), \qquad (2.6.19)$$

$$\Delta P\ =\ \sum_{\mu=0}^{n_d}\ \left(H_s^\mu(q)\ddot{q}\ +\ \dot{q}^T C_s^\mu(q)\dot{q}\ +\ h_s^{G\mu}(q)\right)\Delta\theta_s\,(\mu+1). \qquad (2.6.20)$$

Let us introduce the matrices

$$H_s(q)\ =\ \left[H_s^o(q) \mid H_s^1(q) \mid \cdots \mid H_s^{n_d}(q)\right],$$

$$C_s(q)\ =\ \left[C_s^o(q) \mid C_s^1(q) \mid \cdots \mid C_s^{n_d}(q)\right], \qquad (2.6.21)$$

$$h_s^G(q)\ =\ \left[h_s^{Go}(q) \mid 0 \mid \cdots \mid 0\right].$$

Model (2.6.20) now reduces to

$$\Delta P\ =\ \left(H_s(q)\ddot{q}\ +\ \dot{q}^T C_s(q)\dot{q}\ +\ h_s^G(q)\right)\Delta\theta_s \qquad (2.6.22)$$

and this represents the solution of the stated problem. Model (2.6.22) may, conditionally, be referred to as "sensitivity" model.

Let us note certain properties of model (2.6.22), i.e. of the matrices participating in its construction:

$$H_s^\mu(q),\qquad C_s^\mu(q),\qquad h_s^{G\mu}(q),\qquad \mu\epsilon\{0,1,2,3\}.$$

1) The matrices of model (2.6.22) are independent of the dynamic parameters of mechanism.

2) Model (2.6.22) is applicable to both small and large variations in

parameters.

3) Matrix $H_s^\mu(q)$, $\mu \in \{0,\dots,n_d\}$ is a result of inertial effects. This matrix is symmetric and positive definite.

Proof:

The symmetry of matrices $H_s^\mu(q) \in R^{n \times n}$ directly follows from the form of the matrices (2.6.5) and (2.6.6). Positive definiteness will be proved analogously to the proof of Theorem 1 in 2.4. Let us introduce the notation

$$\vec{p}_{is}^{\,o} = \langle \vec{e}, \vec{r} \rangle_{si} \bar{n}_{is}$$

$$\vec{p}_{is}^{\,\mu} = \left(\vec{e}_i \cdot \vec{q}_{s\mu} \right) \bar{n}_{is} \bar{\xi}_i, \qquad \mu = 1,2,3. \tag{2.6.23}$$

Since $\overline{n_{(i,k)},s} = \bar{n}_{is}\bar{n}_{ks}$, (2.6.5) and (2.6.6) may be written as

$$H_s^\mu(i,k) = \vec{p}_{is}^{\,\mu} \cdot \vec{p}_{ks}^{\,\mu}, \qquad \mu \in \{1,\dots,n_d\} \tag{2.6.24}$$

Expression (2.6.24) is similar to (2.4.3), so analogously to procedure (2.4.3) − (2.4.7), we obtain

$$x^T H_s^\mu(q) x > 0 \tag{2.6.25}$$

where $x \in R^n$ is an arbitrary vector such that $x \neq 0$. This is the proof of positive definiteness.

4) Matrices $C_s^{i\mu}(q) \in R^{n \times n}$ occur as a result of centrifugal and Coriolis effects. These matrices are symmetric, i.e.

$$C_s^{i\mu}(k,\ell) = C_s^{i\mu}(\ell,k), \qquad k,\ell \in N \tag{2.6.26}$$

In addition, the property of antisymmetry also holds

$$C_s^{i\mu}(k,\ell) = -C_s^{k\mu}(i,\ell), \qquad k,i > \ell \tag{2.6.27}$$

Proof:

The property of symmetry with respect to indices k and ℓ (2.6.26) is a direct consequence of the symmetry of matrices $C^i(q, \theta)$ whose elements are given by (2.3.32). The property of antisymmetry will be proved on

96

the basis of (2.6.10):

$$c_s^{k\mu}(i,\ell) = <\vec{e}, \ \vec{r}>_{sk} \cdot \left(\vec{e}_\ell \times <\vec{e}, \ \vec{r}>_{si}\right) \bar{\xi}_\ell \bar{\eta}_{\overline{(k,i)},s}$$

$$= -<\vec{e}, \ \vec{r}>_{si} \cdot \left(\vec{e}_\ell \times <\vec{e}, \ \vec{r}>_{sk}\right) \bar{\xi}_\ell \bar{\eta}_{\overline{(i,k)},s}$$

$$= -c_s^{i\mu}(k,\ell)$$

where $\mu=0$, $i>\ell$. In a similar way, the property of antisymmetry for $\mu=$
$= 1,2,3$ is proved from (2.6.11).

At the end, observe that the construction of model (2.6.22) requires
considerably fewer numerical operations to be performed than the construction of the complete dynamic model. A general flow-chart for the
sensitivity model calculation is given in Fig. 2.11, and a detailed
one in Fig. 2.12.

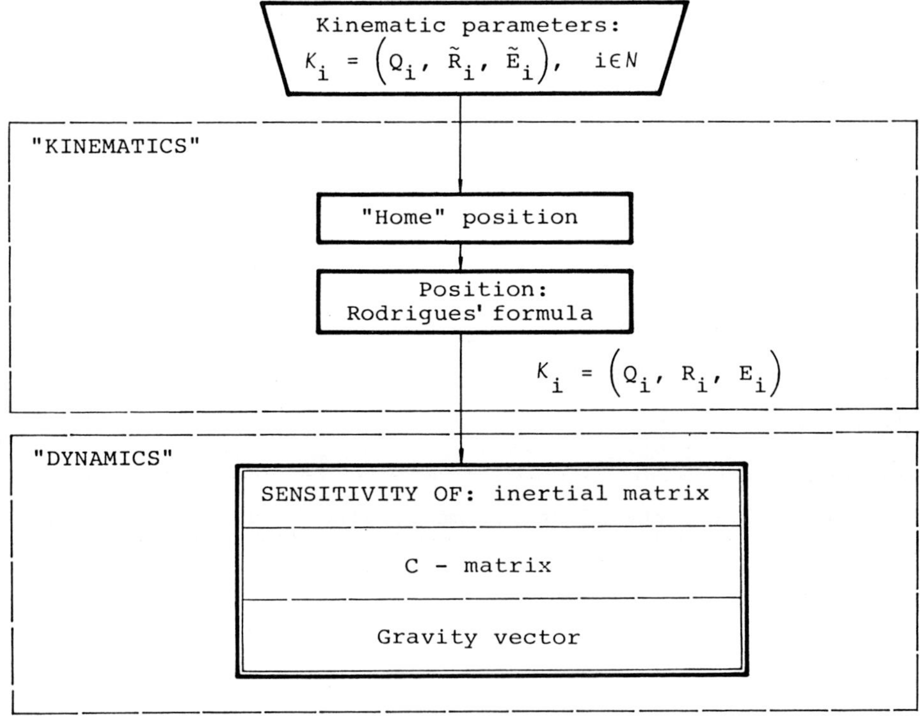

Fig. 2.11. General flow-chart for constructing sensitivity matrices

"Sensitivity of inertial matrix,
C - matrix and gravity vector"

$$H_s^O(i,k) = \langle \vec{e},\ \vec{r}\rangle_{si} \cdot \langle \vec{e},\ \vec{r}\rangle_{sk}\, \bar{\eta}_{\overline{(i,k)},s}$$

$$H_s^\mu(i,k) = \left(\vec{e}_i \cdot \vec{q}_{s\mu}\right)\left(\vec{e}_k \cdot \vec{q}_{s\mu}\right)\bar{\xi}_i \bar{\xi}_k\, \bar{\eta}_{\overline{(i,k)},s}$$

where $i,k \in N$, $\mu \in \{1,2,3\}$.

$$C_s^{io}(k,\ell) = \langle \vec{e},\ \vec{r}\rangle_{si} \cdot \left(\vec{e}_\ell \times \langle \vec{e},\ \vec{r}\rangle_{sk}\right)\bar{\xi}_\ell\, \bar{\eta}_{\overline{(i,k)},s}$$

$$C_s^{i\mu}(k,\ell) = \frac{1}{2}\left[\left(\vec{e}_i \cdot \vec{q}_{s\mu}\right)\vec{\varepsilon}_{\ell k} + \left(\vec{e}_k \cdot \vec{q}_{s\mu}\right)\vec{\varepsilon}_{i\ell} + \right.$$

$$\left. + \left(\vec{e}_\ell \cdot \vec{q}_{s\mu}\right)\vec{\varepsilon}_{ik}\right] \cdot \vec{q}_{s\mu} \bar{\xi}_i \bar{\xi}_k \bar{\xi}_\ell\, \bar{\eta}_{\overline{(i,k)},s}$$

where $i,k,\ell \in N$, $\mu \in \{1,2,3\}$.

$$h_s^G(i) = \langle \vec{e},\ \vec{r}\rangle_{si} \cdot \vec{g}\,\bar{\eta}_{is}$$

where $i \in N$.

Fig. 2.12. Construction of sensitivity matrices (s - the number of the link with variable dynamic parameters)

Appendix 2.1

Kinematic variable $\vec{\beta}_i^O \epsilon W_i$ is under consideration. The procedure for deriving the nonrecursive form of this variable was described in Para. 2.3, and the final result was given by expression (2.3.14). However, a detailed procedure for obtaining (2.3.14) from the complex expression (given in the text in Para. 2.3. immediately before (2.3.14)) for $\vec{\beta}_i^O$ will be presented in this appendix.

According to the derivation given in Para. 2.3, kinematic variable $\vec{\beta}_i^O$ may be represented in the form

$$\vec{\beta}_i^O = \sum_{j=1}^{i} \sum_{k=1}^{j-1} \sum_{\ell=1}^{k-1} \vec{\beta}_{jk\ell}^{(1)} + \sum_{k=1}^{i} \sum_{\ell=1}^{k-1} \vec{\beta}_{k\ell}^{(2)} +$$

$$+ \sum_{j=1}^{i} \sum_{k=1}^{j} \sum_{\ell=1}^{j} \vec{\beta}_{jk\ell}^{(3)} - \sum_{j=1}^{i} \sum_{k=1}^{j-1} \sum_{\ell=1}^{j-1} \vec{\beta}_{jk\ell}^{(4)} \tag{A2.1.1}$$

where

$$\vec{\beta}_{jk\ell}^{(1)} = \vec{\varepsilon}_{\ell k} \times \vec{R}_{j,j-1} \bar{\xi}_k \bar{\xi}_\ell \dot{q}_k \dot{q}_\ell \tag{A2.1.2}$$

$$\vec{\beta}_{k\ell}^{(2)} = \left(\vec{\varepsilon}_{\ell k} \times \vec{r}_{kk} \bar{\xi}_k + 2\vec{\varepsilon}_{\ell k} \xi_k \right) \bar{\xi}_\ell \dot{q}_k \dot{q}_\ell \tag{A2.1.3}$$

$$\vec{\beta}_{jk\ell}^{(3)} = \vec{e}_k \times \left(\vec{e}_\ell \times \vec{r}_{jj} \right) \bar{\xi}_k \bar{\xi}_\ell \dot{q}_k \dot{q}_\ell \tag{A2.1.4}$$

$$\vec{\beta}_{jk\ell}^{(4)} = \vec{e}_k \times \left(\vec{e}_\ell \times \vec{r}_{j-1,j} \right) \bar{\xi}_k \bar{\xi}_\ell \dot{q}_k \dot{q}_\ell \tag{A2.1.5}$$

The problem reduces to transforming form (A2.1.1) into an equivalent form of the type of (2.3.15). To this end, let us consider the following identity

$$\sum_{j=1}^{i} \sum_{k=1}^{j-1} \sum_{\ell=1}^{k-1} \vec{\beta}_{jk\ell}^{(1)} = \sum_{k=1}^{i} \sum_{\ell=1}^{i} \left(\eta_{k\ell} \sum_{j=k+1}^{i} \vec{\beta}_{jk\ell}^{(1)} \right) \tag{A2.1.6}$$

Similarly to deriving expression (2.3.12), let us introduce the following shortened notation for the right-hand side of (A2.1.6)

$$\sum_{k=1}^{i}\sum_{\ell=1}^{i}\eta_{k\ell}\vec{P}_{k\ell} \ , \qquad \vec{P}_{k\ell} = \sum_{j=k+1}^{i}\vec{\beta}_{jk\ell}^{(1)} \ ,$$

and let us note that the following identity holds

$$\sum_{k=1}^{i}\sum_{\ell=1}^{i}\eta_{k\ell}\vec{P}_{k\ell} = \sum_{k=1}^{i}\sum_{\ell=1}^{i}\tfrac{1}{2}\left(\eta_{k\ell}\vec{P}_{k\ell} + \eta_{\ell k}\vec{P}_{\ell k}\right) \qquad (A2.1.7)$$

as well as that the expression under the double sum on the right-hand side of equation is symmetric with respect to indices k and ℓ, i.e. that $\vec{q}_{k\ell} = \vec{q}_{\ell k}$ holds, where

$$\vec{q}_{k\ell} = \tfrac{1}{2}\left(\eta_{k\ell}\vec{P}_{k\ell} + \eta_{\ell k}\vec{P}_{\ell k}\right) = \begin{cases} \dfrac{1}{2}\,\vec{P}_{k\ell} & (k>\ell) \\[2mm] \dfrac{1}{2}\,\vec{P}_{\ell k} & (k<\ell) \end{cases} \qquad (A2.1.8)$$

The property of symmetry permits us to write $\tfrac{1}{2}\vec{P}_{k\ell}$ instead of $\vec{q}_{k\ell}$, stating that $k>\ell$ (for indices $k<\ell$ the order of indices should be changed)

$$\sum_{k=1}^{i}\sum_{\ell=1}^{i}\vec{q}_{k\ell} = \sum_{k=1}^{i}\sum_{\substack{\ell=1 \\ (k>\ell)}}^{i}\tfrac{1}{2}\vec{P}_{k\ell} \qquad (A2.1.9)$$

Thus, (A2.1.7) becomes

$$\sum_{k=1}^{i}\sum_{\ell=1}^{i}\eta_{k\ell}\vec{P}_{k\ell} = \sum_{k=1}^{i}\sum_{\substack{\ell=1 \\ (k>\ell)}}^{i}\tfrac{1}{2}\vec{P}_{k\ell} \qquad (A2.1.10)$$

We will apply this relation to (A2.1.6), taking into account $\vec{\beta}_{jk\ell}^{(1)}$ which is given by (A2.1.2):

$$\sum_{j=1}^{i}\sum_{k=1}^{j-1}\sum_{\ell=1}^{k-1}\beta_{jk\ell}^{(1)} = \sum_{k=1}^{i}\sum_{\substack{\ell=1 \\ (k>\ell)}}^{i}\tfrac{1}{2}\left(\vec{\varepsilon}_{\ell k}\times\sum_{j=k+1}^{i}\vec{R}_{j,j-1}\right)\bar{\bar{\xi}}_{k}\bar{\bar{\xi}}_{\ell}\dot{q}_{k}\dot{q}_{\ell}$$

$$= \sum_{k=1}^{i}\sum_{\substack{\ell=1 \\ (k>\ell)}}^{i}\tfrac{1}{2}\left(\vec{\varepsilon}_{\ell k}\times\vec{R}_{ik}\right)\bar{\bar{\xi}}_{k}\bar{\bar{\xi}}_{\ell}\dot{q}_{k}\dot{q}_{\ell} \qquad (A2.1.11)$$

where $\vec{R}_{ik}$ is the distance vector from the center of mass of the k-th link to the center of mass of the link i.

Let us now consider the second term in (A2.1.1). Similarly to the previous derivation, we obtain

$$\sum_{k=1}^{i}\sum_{\ell=1}^{k-1}\vec{\beta}_{k\ell}^{(2)} = \sum_{k=1}^{i}\sum_{\ell=1}^{i}\eta_{k\ell}\vec{\beta}_{k\ell}^{(2)} = \sum_{k=1}^{i}\sum_{\substack{\ell=1 \\ (k>\ell)}}^{i}\tfrac{1}{2}\vec{\beta}_{k\ell}^{(2)} =$$

$$= \sum_{\substack{k=1 \\ (k>\ell)}}^{i} \sum_{\ell=1}^{i} \left(\frac{1}{2} \vec{\varepsilon}_{\ell k} \times \vec{r}_{kk} \bar{\xi}_k + \vec{\varepsilon}_{\ell k} \xi_k \right) \bar{\xi}_\ell \dot{q}_k \dot{q}_\ell \tag{A2.1.12}$$

The third term of (A2.1.1) may be translated into an equivalent form by using the following identity

$$\sum_{j=1}^{i} \sum_{k=1}^{j} \sum_{\ell=1}^{j} \vec{\beta}_{jk\ell}^{(3)} = \sum_{k=1}^{i} \sum_{\ell=1}^{i} \left(\sum_{j=(k,\ell)}^{i} \vec{\beta}_{jk\ell}^{(3)} \right). \tag{A2.1.13}$$

Since the expression under the double sum on the right-hand side of equation (A2.1.13) is not symmetric with respect to indices k and ℓ, we will use the identity

$$\sum_{k=1}^{i} \sum_{\ell=1}^{i} \vec{p}_{k\ell} = \sum_{k=1}^{i} \sum_{\ell=1}^{i} \frac{1}{2} \left(\vec{p}_{k\ell} + \vec{p}_{\ell k} \right). \tag{A2.1.14}$$

Thus, (A2.1.13) becomes

$$\sum_{j=1}^{i} \sum_{k=1}^{j} \sum_{\ell=1}^{j} \vec{\beta}_{jk\ell}^{(3)} = \sum_{k=1}^{i} \sum_{\ell=1}^{i} \left(\frac{1}{2} \sum_{j=(k,\ell)}^{i} \left(\vec{e}_k \times \left(\vec{e}_\ell \times \vec{r}_{jj} \right) \right. \right.$$
$$\left. \left. + \vec{e}_\ell \times \left(\vec{e}_k \times \vec{r}_{jj} \right) \right) \bar{\xi}_k \bar{\xi}_\ell \dot{q}_k \dot{q}_\ell \right) \tag{A2.1.15}$$

The fourth term in (A2.1.1) may, in a way similar to that for the third, be translated into corresponding equivalent forms

$$\sum_{j=1}^{i} \sum_{k=1}^{j-1} \sum_{\ell=1}^{j-1} \vec{\beta}_{jk\ell}^{(4)} = \sum_{k=1}^{i} \sum_{\ell=1}^{i} \left(\sum_{j=(k+1,\ell+1)}^{i} \vec{\beta}_{jk\ell}^{(4)} \right) =$$
$$= \sum_{k=1}^{i} \sum_{\ell=1}^{i} \left(\frac{1}{2} \sum_{j=(k+1,\ell+1)}^{i} \left(\vec{\beta}_{jk\ell}^{(4)} + \vec{\beta}_{j\ell k}^{(4)} \right) \right) =$$
$$= \sum_{k=1}^{i} \sum_{\ell=1}^{i} \left(\frac{1}{2} \sum_{j=(k+1,\ell+1)}^{i} \left(\vec{e}_k \times \left(\vec{e}_\ell \times \vec{r}_{j-1,j} \right) \right. \right.$$
$$\left. \left. + \vec{e}_\ell \times \left(\vec{e}_k \times \vec{r}_{j-1,j} \right) \right) \bar{\xi}_k \bar{\xi}_\ell \dot{q}_k \dot{q}_\ell \right) \tag{A2.1.16}$$

All four terms on the right-hand side of (A2.1.1) have thus been transformed to a symmetric from with respect to indices k and ℓ. This identity may now be represented in the form

$$\vec{\beta}_i^O = \sum_{k=1}^{i} \sum_{\ell=1}^{i} C_\beta^i(k, \ell)\,\dot{q}_k\dot{q}_\ell \tag{A2.1.17}$$

where $C_\beta^i(k, \ell) = C_\beta^i(\ell, k)$ holds. The expression for $C_\beta^i(k, \ell)$, $k > \ell$ is obtained by substituting (A2.1.11), (A2.1.12), (A2.1.15) and (A2.1.16) into (A2.1.1):

$$
\begin{aligned}
C_\beta^i(k, \ell) = \frac{1}{2} \Big[&\vec{\varepsilon}_{\ell k} \times \vec{R}_{ik}\bar{\xi}_k + \vec{\varepsilon}_{\ell k} \times \vec{r}_{kk}\bar{\xi}_k + 2\vec{\varepsilon}_{\ell k}\xi_k + \\
&+ \sum_{j=k}^{i}\Big[\vec{e}_k \times \big(\vec{e}_\ell \times \vec{r}_{jj}\big) + \vec{e}_\ell \times \big(\vec{e}_k \times \vec{r}_{jj}\big)\Big]\bar{\xi}_k - \\
&- \sum_{j=k+1}^{i}\Big[\vec{e}_k \times \big(\vec{e}_\ell \times \vec{r}_{j-1,j}\big) + \vec{e}_\ell \times \big(\vec{e}_k \times \vec{r}_{j-1,j}\big)\Big]\bar{\xi}_k\Big\}\bar{\xi}_\ell \tag{A2.1.18}
\end{aligned}
$$

Taking into account that

$$\vec{R}_{ik} + \vec{r}_{kk} = \vec{r}_{ik}, \quad \text{and}$$

$$\sum_{j=k}^{i} \vec{r}_{jj} - \sum_{j=k+1}^{i} \vec{r}_{j-1,j} = \vec{r}_{ik},$$

we obtain

$$
\begin{aligned}
C_\beta^i(k, \ell) = \frac{1}{2}\Big[&\vec{\varepsilon}_{\ell k} \times \vec{r}_{ik}\bar{\xi}_k + 2\vec{\varepsilon}_{\ell k}\xi_k + \\
&+ \vec{e}_k \times \big(\vec{e}_\ell \times \vec{r}_{ik}\big)\bar{\xi}_k + \vec{e}_\ell \times \big(\vec{e}_k \times \vec{r}_{ik}\big)\bar{\xi}_k\Big]\xi_\ell \tag{A2.1.19}
\end{aligned}
$$

Since $\vec{\varepsilon}_{\ell k} = \vec{e}_\ell \times \vec{e}_k$ and since the following identity holds

$$\big(\vec{e}_\ell \times \vec{e}_k\big) \times \vec{r}_{ik} + \vec{e}_k \times \big(\vec{e}_\ell \times \vec{r}_{ik}\big) \equiv \vec{e}_\ell \times \big(\vec{e}_k \times \vec{r}_{ik}\big),$$

we obtain

$$
\begin{aligned}
C_\beta^i(k, \ell) &= \Big[\vec{e}_\ell \times \big(\vec{e}_k \times \vec{r}_{ik}\big)\bar{\xi}_k + \vec{\varepsilon}_{\ell k}\xi_k\Big]\bar{\xi}_\ell, \\
&= \vec{e}_\ell \times \Big[\big(\vec{e}_k \times \vec{r}_{ik}\big)\bar{\xi}_k + \vec{e}_k\xi_k\Big]\bar{\xi}_\ell,
\end{aligned}
$$

or, according to the accepted notation $\langle\vec{e}, \vec{r}\rangle_{ik} = \vec{e}_k \times \vec{r}_{ik}\bar{\xi}_k + \vec{e}_k\xi_k$,

$$C_\beta^i(k, \ell) = \vec{e}_\ell \times \langle\vec{e}, \vec{r}\rangle_{ik}\bar{\xi}_\ell \tag{A2.1.20}$$

which represents the solution of the stated problem.

Appendix 2.2

In this appendix we will present a series of equivalent transformations by which variable $\vec{b}_i^{\,O}$ is reduced to the simplest form (2.3.25)-(2.3.26). For this purpose, let us substitute (2.3.20) and (2.3.21) into (2.3.19)

$$\vec{b}_i^{\,O} = - \sum_{k=1}^{i} \sum_{\ell=1}^{k-1} \sum_{\mu=1}^{3} \left(\vec{\varepsilon}_{\ell k} \cdot \vec{q}_{i\mu} \right) \vec{q}_{i\mu} J_{i\mu} \bar{\xi}_k \bar{\xi}_\ell \dot{q}_k \dot{q}_\ell + \sum_{k=1}^{i} \sum_{\ell=1}^{i} \vec{\lambda}_{k\ell}^{\,i} \dot{q}_k \dot{q}_\ell \tag{A2.2.1}$$

Let us consider the first term on the right-hand side of the equality. Applying the same procedure as in Appendix 2.1, we will perform "symmetrization" with respect to indices k and ℓ (see (A2.1.6) - (A2.1.9)). Thus, we obtain

$$\sum_{k=1}^{i} \sum_{\ell=1}^{k-1} \sum_{\mu=1}^{3} \left(\vec{\varepsilon}_{\ell k} \cdot \vec{q}_{i\mu} \right) \vec{q}_{i\mu} J_{i\mu} \bar{\xi}_k \bar{\xi}_\ell \dot{q}_k \dot{q}_\ell =$$

$$= \sum_{\substack{k=1 \\ (k > \ell)}}^{i} \sum_{\ell=1}^{i} \left(\frac{1}{2} \sum_{\mu=1}^{3} \left(\vec{\varepsilon}_{\ell k} \cdot \vec{q}_{i\mu} \right) \vec{q}_{i\mu} J_{i\mu} \bar{\xi}_k \bar{\xi}_\ell \dot{q}_k \dot{q}_\ell \right) \tag{A2.2.2}$$

The second term of (A2.2.1) evidently depends on the form of $\vec{\lambda}_{k\ell}^{\,i}$, which is given by (2.3.22):

$$\vec{\lambda}_{k\ell}^{\,i} = \sum_{\mu=1}^{3} \vec{p}_{k\ell\mu}^{\,i} \left(J_{i[\mu+1]} - J_{i[\mu+2]} \right) \bar{\xi}_k \bar{\xi}_\ell \tag{A2.2.3}$$

where

$$\vec{p}_{k\ell\mu}^{\,i} = \left(\vec{e}_k \cdot \vec{q}_{i[\mu+1]} \right) \left(\vec{e}_\ell \cdot \vec{q}_{i[\mu+2]} \right) \vec{q}_{i\mu} \tag{A2.2.4}$$

By simple trnasformations the sum (A2.2.3) may be translated into the form

$$\vec{\lambda}_{k\ell}^{\,i} = \sum_{\mu=1}^{3} \left(\vec{p}_{k\ell[\mu+2]}^{\,i} - \vec{p}_{k\ell[\mu+1]}^{\,i} \right) J_{i\mu} \bar{\xi}_k \bar{\xi}_\ell \tag{A2.2.5}$$

In addition, it is necessary to transform expression (2.3.21) into a form "symmetric" with respect to indices k and ℓ, which is given by (1.4.4). Analogously to the procedure in Appendix 2.1, we obtain

$$c_\lambda^i(k,\ \ell) = \frac{1}{2}\left(\vec{\lambda}_{k\ell}^{\,i} + \vec{\lambda}_{\ell k}^{\,i}\right), \qquad (i > k,\ \ell) \tag{A2.2.6}$$

where $c_\lambda^i(k,\ \ell) = c_\lambda^i(\ell,\ k)$. Substituting (A2.2.5) into (A2.2.6), we obtain

$$c_\lambda^i(k,\ \ell) = \frac{1}{2}\sum_{\ell=1}^{3}\left[\vec{P}_{k\ell[\mu+2]}^{\,i} - \vec{P}_{k\ell[\mu+1]}^{\,i} + \vec{P}_{\ell k[\mu+2]}^{\,i} - \vec{P}_{\ell k[\mu+1]}^{\,i}\right]J_{i\mu}\bar{\xi}_k\bar{\xi}_\ell \tag{A2.2.7}$$

Let us note that the following holds

$$\vec{P}_{k\ell[\mu+2]}^{\,i} - \vec{P}_{\ell k[\mu+1]}^{\,i} = \left(\vec{e}_k\cdot\vec{q}_{i\mu}\right)\left[\left(\vec{e}_\ell\cdot\vec{q}_{i[\mu+1]}\right)\vec{q}_{i[\mu+2]} - \left(\vec{e}_\ell\cdot\vec{q}_{i[\mu+2]}\right)\vec{q}_{i[\mu+1]}\right] =$$

$$= \left(\vec{e}_k\cdot\vec{q}_{i\mu}\right)\left(\vec{q}_{i\mu}\times\vec{e}_\ell\right) \tag{A2.2.8}$$

since $\vec{q}_{i\mu} = \vec{q}_{i[\mu+1]}\times\vec{q}_{i[\mu+2]}$. In the same way we obtain

$$\vec{P}_{\ell k[\mu+2]}^{\,i} - \vec{P}_{k\ell[\mu+1]}^{\,i} = \left(\vec{e}_\ell\cdot\vec{q}_{i\mu}\right)\left(\vec{q}_{i\mu}\times\vec{e}_k\right) \tag{A2.2.9}$$

Substituting (A2.2.8) and (A2.2.9) into (A2.2.7), we obtain

$$c_\lambda^i(k,\ \ell) = \frac{1}{2}\sum_{\ell=1}^{3}\left[\left(\vec{e}_k\cdot\vec{q}_{i\mu}\right)\left(\vec{q}_{i\mu}\times\vec{e}_\ell\right) + \left(\vec{e}_\ell\cdot\vec{q}_{i\mu}\right)\left(\vec{q}_{i\mu}\times\vec{e}_k\right)\right]J_{i\mu}\bar{\xi}_k\bar{\xi}_\ell \tag{A2.2.10}$$

which is identical to (2.3.24). Finally, let us substitute (A2.2.2) and (A2.2.10) into (A2.2.1):

$$\vec{b}_i^{\,0} = \sum_{k=1}^{i}\sum_{\ell=1}^{i}\left\{\frac{1}{2}\sum_{\mu=1}^{3}\left[\left(\left(\vec{e}_k\times\vec{e}_\ell\right)\cdot\vec{q}_{i\mu}\right)\vec{q}_{i\mu} + \left(\vec{e}_k\cdot\vec{q}_{i\mu}\right)\left(\vec{q}_{i\mu}\times\vec{e}_\ell\right)\right.\right.$$

$$(k > \ell)$$

$$\left.\left. + \left(\vec{e}_\ell\cdot\vec{q}_{i\mu}\right)\left(\vec{q}_{i\mu}\times\vec{e}_k\right)\right]J_{i\mu}\bar{\xi}_k\bar{\xi}_\ell\dot{q}_k\dot{q}_\ell\right\} \tag{A2.2.11}$$

wherefrom (2.3.26) is directly obtained. The simplest closed-form expression for dynamic coefficient $\vec{b}_i^{\,0}$ has thus been determined.

Chapter 3
Computer-Aided Generation of Numeric-Symbolic Robot Model

3.1. Introduction

In the previous chapter we considered some algorithms for mathematical modelling of robot manipulators. We developed a closed-form dynamic robot model which is very suitable for program implementation. Further, we indicated how such a model could be used to prove the property of symmetry and positive-definiteness of inertial matrix, and antisymmetry of C-matrices. We also resolved the problem of linearized and sensitivity model generation. In this chapter, we wish to consider the computer implementation of the developed algorithms.

In Ch. 1 we indicated that there exist 3 fundamental methods in program implementation, which could be referred to as numeric, symbolic and numeric-symbolic (analytic) methods. The classification of these methods is performed according to the treatment of the variables which take part in modelling procedures. Both kinematic and dynamic variables depend generally on parameters, joint coordinates and joint velocities. We observe that the variables represent scalar or vector functions. They can be treated as

1) numerical quantities (computed for given parameters, joint coordinates and velocities) and

2) functions of parameters and generalized coordinates.

The approach 1) is the numerical modelling method described in Ch. 1. Such a method is often burdened with a large number of floating-point multiplications, and thus difficult for real-time implementation. In the case 2) a basic challenge, however, is that of resolving the following problems:

a) how to choose the arguments from the standpoint of computer memory consumption and speed of operation,

b) how to describe the dependence of an arbitrary variable on arguments in an uniform way,

c) how to realize the algebraic operations between the functions, and

d) how to perform the optimization of functions in order to reduce the number of floating-point multiplications/additions (for example, functions $z_1 = (\cos^2 x + \sin^2 x)y$ and $z_2 = y$ are equivalent, but the structure of z_2 is obviously simpler).

In this chapter we shall consider all these problems. The first problem is rather delicate and imposes a significant preliminary analysis. As explained in Ch. 1, any parameter, joint coordinate and its derivative could be considered as an argument. The result is the symbolic modelling approach, which employs non-numeric quantities, i.e. symbols. Nevertheless, the number of parameters describing kinematic and dynamic properties of robot links might be rather large. For instance, we indicated in Ch. 1 that a 6 degree-of-freedom manipulator is described by about 100 parameters. Adding 12 joint coordinates/velocities, and having in mind the complexity of nonlinear algebraic operations throughout the modelling procedure, it follows that this method might be rather complicated.

We see that the symbolic approach may result in many troubles concerning particularly software complexity. On the other hand, for a given robot, the parameters are usually taken to be constants. Thus, the number of arguments may be reduced to 2n (n-number of joints/links) if we accept only joint coordinates and velocities as arguments. It has been shown that even in this case the analytic structure of the various expressions describing Coriolis and centrifugal effects is rather complex. For this reason, it is the most elegant and useful to take only joint coordinates as arguments. The closed-form model, described in Para. 2.3, is very convenient for the last approach, because the dynamic model matrices

$$H(q), \ C(q) \quad \text{and} \quad h_G(q)$$

depend only on joint coordinates.

The last discussion indicates that the most suitable method is obviously the numeric-symbolic one. That is, the parameters are taken to be constants, whereas the joint coordinates are treated as symbols (arguments).

The problem b) related to the analytic structure of functions-variables
will be discussed in Para. 3.2. It will be shown that any kinematic
or dynamic variable can be represented by a unifold sequence of data.
This unifold form will be obtained by introducing a specific nonlinear
transformation from the joint-coordinate space $Q \subset R^n$ to a newly defined
space $X \subset R^{3n}$ of arguments. This will yield the transformation of complex
trigonometric functions to polynomials. As indicated in Ch. 1, these
polynomials can be described by 2 sets of numeric data

    - vector of coefficients, and

    - matrix of exponents.

This pair uniquely defines the polynomial, i.e. the variable. We shall
call this pair the polynomial matrix. This matrix describes the analytic structure of a model variable. Further, we will prove a rather important theorem stating that the elements of E-matrices correspond to
the set $\{0, 1, 2\}$. This is significant from the standpoint of memory
consumption.

In Para 3.3 the algebra of polynomial matrices will be developed.
We shall show that the space of polynomial matrices constitutes a vector space. Naturally, we shall introduce the addition, multiplication
by a scalar, cross and dot product etc. It will be shown that these
operations are suitable for programming on a digital computer.

We shall then try to resolve one of the fundamental questions related
to the unifoldness of polynomial matrices. As we shall see, there is
an infinite number of polynomial matrices representing the same variable. This is essentially the consequence of the expansion of the space
of arguments $Q \subset R^n$ (joint coordinates) to the space $X \subset R^{3n}$. The mapping
$Q \rightarrow X$ leads to the appearance of nonlinear relations between arguments
of X space. It implies that there exist different, but equivalent,
polynomial matrices. Now, we encounter the problem of how to determine
a unique representative of the set of equivalent polynomial matrices.
It will be shown that there exists such a representative which corresponds to the polynomial matrix with a minimal number of rows. An algorithm which reduces any polynomial matrix to one with a minimal number of rows will be developed.

The second part of this chapter will be devoted to the application of the obtained results to robot modelling. In Para. 3.5 we shall present an algorithm for nonlinear dynamic model construction in analytic form. Further, two intrinsically different algorithms for linearized robot model construction will be developed. The first one is developed using equations given in Para. 2.5. The second one is based on the theorem about partial derivatives of polynomial matrices (see Para. 3.6). This algorithm turns out to be quite elegant and useful for applications. In the text to follow, we shall approach the problem of analytic sensitivity model derivation. Then, we shall treat a very practical question concerning the problem of construction of approximative robot models in a systematic fashion. Consistent with this idea, we followed the approach based on imposing some predefined error-criterion. The developed algorithm answers the question about the terms of polynomials, i.e. rows in corresponding polynomial matrices, which could be neglected under the condition of nonviolating the predefined error. At the end of this chapter an example showing the fundamental importance of approximative models from the standpoint of numerical complexity reduction and real-time application, will be presented.

3.2. Numeric-Symbolic Representation of Variables

Consider a robot manipulator with n links and n joints. The set of kinematic variables is defined in Para. 2.2 as

$$K_i = \left(Q_i, R_i, E_i \right), \qquad i \in N = \{1, \ldots, n\},$$

where

$$Q_i = \left(\vec{q}_{i1}, \vec{q}_{i2}, \vec{q}_{i3} \right) \quad \text{- local coordinate frame attached to link } i,$$

$$R_i = \left\{ \vec{r}_{ii}, \vec{r}_{i,i+1} \right\} \quad \text{- set of distance vectors from joints to mass centers,}$$

$$E_i = \left\{ \vec{e}_i \right\} \quad \text{- set of joint axes unit vectors,}$$

for simple kinematic chains connected by revolute and sliding joints.

Following the closed-form model equations in Para. 2.3 (Fig. 2.7) we see that the variables (vector and scalar expressions) which figure in model equations are obtained from the elements of set K_i and $D_i = \left(m_i, \right.$

$$J_{i1},\ J_{i2},\ J_{i3}\Big),\ i \in N.$$

Let us introduce now the notation of "mechanism variable", or simply "variable":

Definition

Mechanism variables are

1) elements of K_i, $\forall i \in N$,

2) variables obtained by linear operations in K_i, $\forall i \in N$, and

3) variables obtained by nonlinear operations in K_i, $\forall i \in N$.

The linear operations mentioned above represent the addition and substraction of vectors, while the nonlinear operations represent (vector and scalar) multiplication of vectors as well as multiplication by a scalar. For example, a variable from the set 2) is

$$\vec{r}_{ij} = \vec{r}_{jj} - \vec{r}_{j,j+1} + \vec{r}_{j+1,j+1} - \cdots + \vec{r}_{ii}, \qquad (i>j),$$

and from the set 3)

$$\vec{\varepsilon}_{ij} = \vec{e}_i \times \vec{e}_j$$

$$<\vec{e},\ \vec{r}>_{ij} = \vec{e}_j \times \vec{r}_{ij}\bar{\xi}_j + \vec{e}_j \xi_j \qquad\qquad (3.2.1)$$

$$p_{ij\mu} = \vec{e}_i \cdot \vec{q}_{j\mu}.$$

Variables of a mechanism can be described in 3 ways: numerically, symbolically and in a mixed way (combining numerical and symbolical approach). Numerical representative of a variable represents the set of reals determining its value for a given vector of joint coordinates. Symbolic representative of a variable is composed of mathematical symbols which define its dependence on parameters and joint coordinates. The third type of variable description may be referred to as numeric - symbolic or analytic. Here, parameters are treated numerically, and the joint coordinates - symbolically. As explained in Ch. 1, it is very useful to develop a modelling algorithm based on analytical description of variables.

*Numeric-symbolic representation of
variables*

Consider the set of unit vectors of local coordinate frames Q_i, $i \in N$.
The basic equation determining these vectors is the finite-rotation
formula (2.2.4) which can be presented as

$$\vec{q}_{ij} = \begin{bmatrix} \vec{q}_{ij}^{(3)} & \vec{q}_{ij}^{(2)} & \vec{q}_{ij}^{(1)} \end{bmatrix} \begin{bmatrix} \cos q_i \\ \sin q_i \\ 1 \end{bmatrix}, \quad j=1,2,3, \quad\quad (3.2.2)$$

with

$$\vec{q}_{ij}^{(3)} = \vec{e}_i \times \left(\vec{q}_{ij}^{o} \times \vec{e}_i \right) \bar{\xi}_i$$

$$\vec{q}_{ij}^{(2)} = \vec{e}_i \times \vec{q}_{ij}^{o} \bar{\xi}_i \quad\quad (3.2.3)$$

$$\vec{q}_{ij}^{(1)} = \left(\vec{e}_i \cdot \vec{q}_{ij}^{o} \right) \vec{e}_i \xi_i + \vec{q}_{ij}^{o} \xi_i.$$

Given the joint coordinates $q_1, \ldots, q_{i-1}$, we can determine the vectors $\vec{e}_i$
and $\vec{q}_{ij}^{o}$. Let us determine their analytic dependence on q_{i-1}. From
(2.2.5) we have

$$\vec{e}_i = Q_{i-1} \tilde{e}_i \quad\quad (3.2.4)$$

where $\tilde{e}_i$ denotes the unit vector of joint i with respect to the coordinate frame of link i-1. We also have

$$\vec{q}_{ij}^{o} = Q_{i-1} \tilde{q}_{ij}^{o} \qu\quad (3.2.5)$$

with $\tilde{q}_{ij}^{o}$ given in (i-1)-th coordinate frame. From (3.2.4) and (3.2.5)
it follows that

$$\vec{q}_{ij}^{(k)} = Q_{i-1} \tilde{q}_{ij}^{(k)}, \quad (k=1,2,3) \qu\quad (3.2.6)$$

Substituting equations (3.2.6) into (3.2.2), we obtain

$$\vec{q}_{ij} = Q_{i-1} \begin{bmatrix} \tilde{q}_{ij}^{(3)} & \tilde{q}_{ij}^{(2)} & \tilde{q}_{ij}^{(1)} \end{bmatrix} \begin{bmatrix} \cos q_i \\ \sin q_i \\ 1 \end{bmatrix} \qu\quad (3.2.7)$$

Having in mind that $Q_{i-1} = \left[\vec{q}_{i-1,1} \ \vec{q}_{i-1,2} \ \vec{q}_{i-1,3}\right]$ and

$$\vec{q}_{i-1,j} = \left[\vec{q}_{i-1,j}^{(3)} \ \vec{q}_{i-1,j}^{(2)} \ \vec{q}_{i-1,j}^{(1)}\right] \begin{bmatrix} \cos q_{i-1} \\ \sin q_{i-1} \\ 1 \end{bmatrix} \qquad (3.2.8)$$

we obtain by substituting in (3.2.7)

$$\vec{q}_{ij} = \left[\vec{q}_{ij}^{(9)} \ \cdots \ \vec{q}_{ij}^{(1)}\right] \begin{bmatrix} \cos q_{i-1} \cos q_i \\ \cos q_{i-1} \sin q_i \\ \cos q_{i-1} \\ \hline \sin q_{i-1} \cos q_i \\ \sin q_{i-1} \sin q_i \\ \sin q_{i-1} \\ \hline \cos q_i \\ \sin q_i \\ 1 \end{bmatrix} \qquad (3.2.9)$$

It turns out that the elements of the column-vector on the right hand side of (3.2.9) coincide with the elements of the Cartesian product

$$\left(\cos q_{i-1}, \ \sin q_{i-1}, \ 1\right) \times \left(\cos q_i, \ \sin q_i, \ 1\right). \qquad (3.2.10)$$

This vector will be shortly denoted as $\mathrm{Rot}\left(q_{i-1}, q_i\right)$. Thus, equations (3.2.2) and (3.2.9) are reduced to

$$\vec{q}_{ij} = Q_{ij}^{(3)} \ \mathrm{Rot} \ q_i = Q_{ij}^{(9)} \ \mathrm{Rot}\left(q_{i-1}, q_i\right) \qquad (3.2.11)$$

where $Q_{ij}^{(3)} \in R^{3\times 3}$ and $Q_{ij}^{(9)} \in R^{3\times 9}$ correspond to matrices $\left[\vec{q}_{ij}^{(3)} \ \vec{q}_{ij}^{(2)} \ \vec{q}_{ij}^{(1)}\right]$ and $\left[\vec{q}_{ij}^{(9)} \ \cdots \ \vec{q}_{ij}^{(1)}\right]$, respectively.

Obviously, the general form of (3.2.11) is

$$\vec{q}_{ij} = Q_{ij}^{(3^i)} \ \mathrm{Rot}\left(q_1, \ldots, q_i\right) \qquad (3.2.12)$$

Matrix $Q_{ij}^{(3^i)} \in R^{3\times 3^i}$ is independent of joint coordinates. It depends on kinematic parameters. Vector $\mathrm{Rot}\left(q_1, \ldots, q_i\right) \in R^{3^i}$ consists of the elements of Cartesian product

$$\left(\cos q_1, \sin q_1, 1\right) \times \cdots \times \left(\cos q_i, \sin q_i, 1\right) \qquad (3.2.13)$$

Now, we can easily expain the difference between symbolical and numerical-symbolic (analytical) representations of variables. In both representations it is assumed that joint coordinates $q_1, \ldots, q_i$ are unknown. Thus, the elements of $\mathrm{Rot}\left(q_1, \ldots, q_i\right)$ cannot be substituted by real numbers. They must remain symbolic. The basic difference is in matrix $Q_{ij}^{(3^i)}$. If one accepts that the elements of $Q_{ij}^{(3^i)}$ represent functions of parameters, then a symbolic variable representation is assumed. If the elements represent real numbers calculated on the basis of known parameters of a given manipulator, then we approach the numerical-symbolic form of a variable. Let us consider the remaining elements of the set of kinematic variables K_i: $R_i = \left\{\vec{r}_{ii}, \vec{r}_{i,i+1}\right\}$ and $E_i = \left\{\vec{e}_i\right\}$ given as

$$\vec{r}_{ii} = Q_i \tilde{r}_{ii} + q_i \vec{e}_i \xi_i$$

$$\vec{r}_{i,i+1} = Q_i \tilde{r}_{i,i+1} \qquad (3.2.14)$$

$$\vec{e}_{i+1} = Q_i \tilde{e}_{i+1}$$

Substituting Equation (3.2.12) into matrix $Q_i = \left[\vec{q}_{i1} \ \vec{q}_{i2} \ \vec{q}_{i3}\right]$, we get

$$Q_i = \left[Q_{i1}^{(3^i)} \ \vdots \ Q_{i2}^{(3^i)} \ \vdots \ Q_{i3}^{(3^i)}\right] \mathrm{Rot}\left(q_1, \ldots, q_i\right). \qquad (3.2.15)$$

Substituting into (3.2.14), it follows

$$\vec{r}_{ii} = R_{ii}^{(3^i)} \mathrm{Rot}\left(q_1, \ldots, q_i\right) + q_i \vec{e}_i \xi_i$$

$$\vec{r}_{i,i+1} = R_{i,i+1}^{(3^i)} \mathrm{Rot}\left(q_1, \ldots, q_i\right) \qquad (3.2.16)$$

$$\vec{e}_{i+1} = E_{i+1}^{(3^i)} \mathrm{Rot}\left(q_1, \ldots, q_i\right)$$

The first of these equations shows that the form

$$\vec{v}_i = V_i^{(3^i)} \mathrm{Rot}\left(q_1, \ldots, q_i\right) \qquad (3.2.17)$$

is not general enough. Besides, the variables obtained by nonlinear operations in K_i cannot be represented by the form (3.2.17), too. For example, the scalar product $\vec{v}_i \cdot \vec{v}_i$ cannot be represented by (3.2.17), because it includes a series of functions which do not figure in $\text{Rot}\left(q_1,\ldots,q_i\right)$. Thus, we will introduce a more general form:

$$v = V^{(m)} \begin{bmatrix} \left(\cos q_1\right)^{c_{11}}\cdots\left(\cos q_n\right)^{c_{n1}} \left(\sin q_1\right)^{s_{11}}\cdots\left(\sin q_n\right)^{s_{n1}} q_1^{u_{11}}\cdots q_n^{u_{n1}} \\ \vdots \\ \left(\cos q_1\right)^{c_{1m}}\cdots\left(\cos q_n\right)^{c_{nm}} \left(\sin q_1\right)^{s_{1m}}\cdots\left(\sin q_n\right)^{s_{nm}} q_1^{u_{1m}}\cdots q_n^{u_{nm}} \end{bmatrix}$$

$$(3.2.18)$$

where

$V^{(m)}$ — is $\ell \times m$ constant matrix ($\ell=3$ if the variable v is a 3D vector, and $\ell=1$ for a scalar)

c_{ij}, s_{ij}, u_{ij} ($i=1,\ldots,n$; $j=1,\ldots,m$) — are exponents of cosines, sines and joint coordinates, respectively.

Obviously, when $m=3^i$ the form (3.2.18) can easily be reduced to (3.2.17). Notice that m is not fixed as the dimension of $\text{Rot}\left(q_1,\ldots,q_i\right)$. Generally, $m<3^i$ holds for the vectors representable by (3.2.17). The dimension 3^i enables to comprise all possible combinations of cosines and sines of joint coordinates.

The analytical form (3.2.18) may be represented as

$$v = \sum_{k=1}^{m} V_k \left(\cos q_1\right)^{c_{1k}}\cdots\left(\cos q_n\right)^{c_{nk}}\left(\sin q_1\right)^{s_{1k}}\cdots\left(\sin q_n\right)^{s_{nk}} q_1^{u_{1k}}\cdots q_n^{u_{nk}}$$

$$(3.2.19)$$

where V_k is the k-th column of the matrix $V^{(m)}$. Let us introduce the vector of new arguments

$$\left(x_1,\ldots,x_N\right)=\left(\cos q_1,\ldots,\cos q_n,\ \sin q_1,\ldots,\sin q_n,\ q_1,\ldots,q_n\right) \qquad (3.2.20)$$

and the vector of exponents

$$\left(\varepsilon_{1k},\ldots,\varepsilon_{Nk}\right)=\left(c_{1k},\ldots,c_{nk},\ s_{1k},\ldots,s_{nk},\ u_{1k},\ldots,u_{nk}\right) \qquad (3.2.21)$$

where, obviously, ε_{ik} ($i=1,\ldots,N$) represents a positive integer or zero. Substituting (3.2.20) and (3.2.21) into (3.2.19), we obtain

$$v = \sum_{k=1}^{m} V_k x_1^{\varepsilon_{1k}} \cdots x_N^{\varepsilon_{Nk}} \qquad N = 3n. \qquad (3.2.22)$$

This form will be referred to as a polynomial numerical-symbolic form or, simply, a polynomial.

If we were to implement the forms (3.2.22) on a digital computer, then we would have to transform them into some suitable numerical forms. For this reason we will introduce matrix representatives of polynomials (3.2.22), which will be referred to as "polynomial matrices".

Definition of a polynomial matrix

The polynomial matrix of a variable v given by the polynomial (3.2.22) represents the pair

$$S_v^{(m)} = \left(v^{(m)^T}, E_v \right)$$

where T denotes the transposition of vector, $v^{(m)^T} = \begin{bmatrix} v_1^T \\ \vdots \\ v_m^T \end{bmatrix}$ with V_i defined in (3.2.22),

E_v denotes the matrix of exponents

$$E_v = \left[\begin{array}{ccc|ccc|ccc} c_{11} & \cdots & c_{n1} & s_{11} & \cdots & s_{n1} & u_{11} & \cdots & u_{n1} \\ \vdots & & & & & & & & \\ c_{1m} & \cdots & c_{nm} & s_{1m} & \cdots & s_{nm} & u_{1m} & \cdots & u_{nm} \end{array} \right] \qquad (3.2.23)$$

We see from the definition that the polynomial matrix completely defines the coefficients and exponents of the corresponding polynomial.

A very important property of polynomial matrices of kinematic and dynamic variables can be formulated by the following theorem:

Theorem

If v represents an arbitrary variable of a robotic manipulator, then the elements of the corresponding matrix of exponents E_v belong to the set $\{0, 1, 2\}$. In the case, of a kinematic variable, the elements of E_v belong to the set $\{0, 1\}$.

This theorem will be precisely proved in Appendix 3.1.

3.3. Algebra of Polynomial Matrices

As we have seen in the previous paragraph, any model variable of a robot manipulator can be represented by the polynomial

$$v = \sum_{k=1}^{m} v_k x_1^{\varepsilon_{1k}} \cdots x_N^{\varepsilon_{Nk}}, \qquad N = 3n,$$

where arguments x_i and x_{n+i} are interdependent

$$x_i^2 + x_{n+i}^2 = 1 \quad \text{for} \quad i \in N \tag{3.3.1}$$

This variable can also be represented by the polynomial matrix

$$S_v^{(m)} = \left(\begin{bmatrix} v_1^T \\ \vdots \\ v_m^T \end{bmatrix}, \begin{bmatrix} \varepsilon_{11} & \cdots & \varepsilon_{N1} \\ \vdots & & \vdots \\ \varepsilon_{1m} & \cdots & \varepsilon_{Nm} \end{bmatrix} \right) \tag{3.3.2}$$

Let us prove a theorem which describes the mathematical operations between polynomial matrices.

Theorem

The set of polynomial matrices corresponding to vector variables of a robot manipulator constitutes the vector space.

Proof

Let $V = \left\{ S_v^{(m)} \mid v \in R^3, m \in N \right\}$ be the set of polynomial matrices of vectors $v \in R^3$. We denote the function, mapping $V \times V$ into V, by G_+. We suppose that

$$\left(S_1, S_2 \right) \in V \times V$$

where S_i designates $S_{v_i}^{(m_i)}$. Let us introduce the addition as

$$S_{V_1}^{(m_1)} + S_{V_2}^{(m_2)} = \left(\left[\frac{V^{(m_1)^T}}{V^{(m_2)^T}} \right] , \left[\frac{E_{V_1}}{E_{V_2}} \right] \right) . \tag{3.3.3}$$

We also define the function G. which maps R×V into V. If $(r, S) \in R \times V$ then we write $r \cdot S$ for the element G. $((r, S))$ of the set V, i.e.

$$r S_V^{(m)} = G. \left(\left(r, S_V^{(m)} \right) \right).$$

The quantity $r S_V^{(m)}$ is said to be product of r and $S_V^{(m)}$. The function G. is defined as

$$r S_V^{(m)} = \left(\left[r V^{(m)^T} \right], E_V \right) \tag{3.3.4}$$

with r being a real number. When r represents a symbolic variable x_i, the product is defined as

$$x_i S_V^{(m)} = \left(V^{(m)^T}, \begin{bmatrix} \varepsilon_{11} & \cdots & \varepsilon_{i-1,1} & \varepsilon_{i1}^{+1} & \varepsilon_{i+1,1} & \cdots & \varepsilon_{N1} \\ \vdots & & \vdots & \vdots & \vdots & & \vdots \\ \varepsilon_{1m} & \cdots & \varepsilon_{i-1,m} & \varepsilon_{im}^{+1} & \varepsilon_{i+1,m} & \cdots & \varepsilon_{Nm} \end{bmatrix} \right) \tag{3.3.5}$$

with $i \in \{1, \ldots, N\}$.

The set V corresponds to a vector space if the relation G_+ satisfies the following conditions

1) $S_1 + S_2 = S_2 + S_1$ for any S_1 and S_2 in V,

2) $S_1 + \left(S_2 + S_3 \right) = \left(S_1 + S_2 \right) + S_3$ for any S_1, S_2, S_3 in V,

3) There exists a unique element 0 in V such that $S + 0 = 0 + S = S$ for any S in V, and

4) For any S in V, there is a unique element $-S$ in V such that $S + (-S) = = (-S) + S = 0$.

Using the definition (3.3.3) of the function G_+, the properties 1) and 2) can easily be proved. The 0-element is

$$S_o^{(1)} = ([0 \quad 0 \quad 0], [0 \cdots 0]).$$

and the inverse element $-S$

$$-S_v^{(m)} = \left(-V^{(m)^T}, E_v\right).$$

The problem of polynomial matrices uniqueness will be considered in the following section.

The function G. should satisfy the following conditions:

1) $r_1 \cdot \left(r_2 \cdot S\right) = \left(r_1 r_2\right) \cdot S$ for any r_1, r_2 in R and any S in V,

2) $1 \cdot S = S$ for any S in V,

3) $r \cdot \left(S_1 + S_2\right) = r \cdot S_1 + r \cdot S_2$ for any r in R and any S_1 and S_2 in V, and

4) $\left(r_1 + r_2\right) \cdot S = r_1 \cdot S + r_2 \cdot S$ for any r_1, r_2 in R and any S in V.

We can easily prove these properties using the definition (3.3.4) and (3.3.5) of G. In the text to follow, instead of $r \cdot S$ we shall simply write rS. This accomplishes the proof.

Let us introduce the scalar and vector products for the polynomial matrices. The vector product is defined as

$$S_{v_1}^{(m_1)} \times S_{v_2}^{(m_2)} = \sum_{k=1}^{m_1} \sum_{\ell=1}^{m_2} \left(\vec{v}_{1k} \times \vec{v}_{2\ell}\right) x_1^{\varepsilon_{1k}^1 + \varepsilon_{1\ell}^2} \cdots x_N^{\varepsilon_{Nk}^1 + \varepsilon_{N\ell}^2} \tag{3.3.6}$$

with ε_{ik}^1 belonging to matrix of exponents E_{v_1}, and $\varepsilon_{i\ell}^2$ to matrix E_{v_2}. Using polynomial matrices, the polynomial (3.3.6) can be represented as

$$\left(V^{(m_1)^T}, E_{v_1}\right) \times \left(V^{(m_2)^T}, E_{v_2}\right) = \left(\left(V^{(m_1)} \times V^{(m_2)}\right)^T, E_{v_1} + E_{v_2}\right) \tag{3.3.7}$$

where $V^{(m_1)} \times V^{(m_2)} \in R^{3 \times (m_1 m_2)}$ is the matrix whose columns are vectors $\vec{v}_{1k} \times \vec{v}_{2\ell}$ $(k=1,\ldots,m_1;\ \ell=1,\ldots,m_2)$, and $E_{v_1} + E_{v_2}$ – matrix whose rows are $\left[\varepsilon_{1k}^1 + \varepsilon_{1\ell}^2 \cdots \varepsilon_{Nk}^1 + \varepsilon_{N\ell}^2\right]$.

Similarly, we define the scalar product $S_1 \cdot S_2$:

$$\left(V^{(m_1)^T},\ E_{v_1}\right)\cdot\left(V^{(m_2)^T},\ E_{v_2}\right) = \left(\left(V^{(m_1)}\cdot V^{(m_2)}\right)^T,\ E_{v_1}+E_{v_2}\right), \quad (3.3.8)$$

as well as the product of a polynomial matrix for a scalar variable s
and a polynomial matrix for a vector variable v:

$$\left(V_s^{(m_1)^T},\ E_s\right)\left(V_v^{(m_2)^T},\ E_v\right) = \left(\left(V_s^{(m_1)}\ V_v^{(m_2)}\right)^T,\ E_s+E_v\right). \quad (3.3.9)$$

Example

All established operations can easily be applied to any expression which
constitutes a robot modelling algorithm. For example, the finite rotation formula (2.2.4) in analytic form becomes

$$S_{q_{ij}} = S_{e_i} \times S_q x_i \bar{\xi}_i + S_q x_{n+1} \bar{\xi}_i + \left(S_{e_i}\cdot S_{q_{ij}^o}\right) S_{e_i} \bar{\xi}_i + S_{q_{ij}^o} \xi_i \quad (3.3.10)$$

with $S_q = S_{q_{ij}^o} \times S_{e_i}$.

3.4. Optimization of Polynomial Matrices

Consider a variable v in the polynomial form

$$v = \sum_{k=1}^{m} V_k\ x_1^{\varepsilon_{1k}} \cdots x_N^{\varepsilon_{Nk}}$$

where $N = 3n$ and

$$x_i = \cos q_i$$

$$x_{n+i} = \sin q_i \qquad\qquad (3.4.1)$$

$$x_{2n+i} = q_i$$

for $i\in N$. The polynomial arguments x_j, $j=\{1,\ldots,3n\}$ are obviously interdependent. For example, the well known trigonometric relations between
sine and cosine functions cause the following identities

$$x_i^2 = 1 - x_{n+i}^2$$

$$x_i^4 = 1 - 2x_{n+i}^2 + x_{n+i}^4.$$

(3.4.2)

On the other hand, let us consider the polynomial

$$w = \sum_{k=1}^{m+1} w_k \, x_1^{\xi_{1k}} \cdots x_N^{\xi_{Nk}}$$

and assume that

$$w_k = v_k, \qquad k \in \{1,\ldots,m\}$$

$$w_{m+1} = w_m$$

$$\xi_{jk} = \varepsilon_{jk}, \qquad j \in \{1,\ldots,N\}, \qquad k \in \{1,\ldots,m-1\}$$

$$\xi_{jm} = \varepsilon_{jm}, \qquad j \in \{1,\ldots,i-1,\ i+1,\ldots,N\}; \qquad \varepsilon_{im} = \varepsilon_{im}+2,$$

$$\xi_{j,m+1} = \varepsilon_{jm}, \qquad j \in \{1,\ldots,n+i-1,\ n+i+1,\ldots,N\}; \qquad \varepsilon_{n+i,m} = \varepsilon_{n+i,m}+2.$$

Taking into account the first of the relations (3.4.2), we see that the
polynomials v and w are identical. The corresponding polynomial matrices are also equivalent, i.e.

$$S_v^{(m)} = S_w^{(m+1)},$$

$$\left(V_v^{(m)^T}, \ E_v \right) = \left(V_w^{(m+1)^T}, \ E_w \right).$$

It follows that we can present the variable v by 2 different polynomial matrices. By the use of mathematical induction one can derive the
following conclusion: For an arbitrary number $m_\ell > m$, there exists v_ℓ,
such that

$$\left(V_v^{(m)^T}, \ E_v \right) = \left(V_{v_\ell}^{(m_\ell)^T}, \ E_{v_\ell} \right).$$

(3.4.3)

That is, there is an unlimited number of different polynomial matrices
corresponding to the same variable v. The polynomial matrices

$$S_{v_\ell}^{(m_\ell)} = \left(V_{v_\ell}^{(m_\ell)^T}, \ E_{v_\ell} \right)$$

(3.4.4)

are said to be equivalent. The above matrix is m_ℓ - dimensional, but it would be especially valuable to determine a minimal-dimension polynomial matrix corresponding to the variable v:

$$\underline{S}_v^{(m)} = \min_{m_\ell}\left(V_{v_\ell}^{(m_\ell)^T}, E_{v_\ell}\right) \qquad (3.4.5)$$

The polynomial corresponding to this matrix will be called the minimal form of variable v. This form can be considered to be unique for a given variable v if the following 3 conditions are satisfied:

1) The are no 2 rows in $V_v^{(m)^T}$ which are zero-vectors.

2) There are no identical rows in E_v.

3) It is not possible to find any 2 rows which can be united into one due to the identity $x_i^2 + x_{n+i}^2 = 1$ for any subscript i. In other words, there are no 2 rows whose monomials $M_k = v_k x_1^{\varepsilon_{k1}} \cdots x_N^{\varepsilon_{kN}}$ and $M_j = v_j x_1^{\varepsilon_{j1}} \cdots x_N^{\varepsilon_{jN}}$ satisfy the identity

$$M_k + M_j = v_k\, x_1^{\varepsilon_{\ell 1}} \cdots x_N^{\varepsilon_{\ell N}}\left(x_i^2 + x_{n+i}^2\right) = v_k\, x_1^{\varepsilon_{\ell 1}} \cdots x_N^{\varepsilon_{\ell N}}$$

with $v_k = v_j$.

If the condition (1) were not satisfied, the zero-rows of matrix $V_v^{(m)^T}$, as well as the corresponding row in matrix E_v^m, could be rejected, and this reduces the dimension m. If the condition 2) were not satisfied, i.e. if there exist 2 identical rows in matrix E_v, the corresponding rows in V_v^m might be added one to the other, and this reduces the dimension by 1.

The last conclusions show that the polynomial matrix corresponding to the minimal form is the unique representative of a variable v. In the text to follow, the notion of the vector space of polynomial matrices will mean the space of minimal forms. In fact, this vector space satisfies the properties of unique neutral and inverse element (see Para. 3.3). The procedure for determining the minimal form starting from an arbitrary polynomial matrix is illustrated in Fig. 3.1.

120

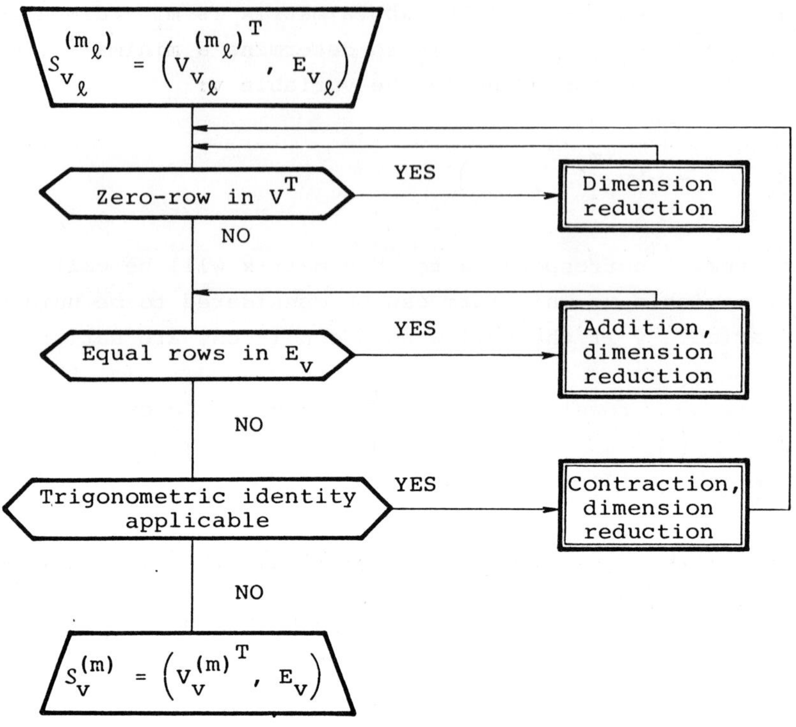

Fig. 3.1. Minimization of a polynomial matrix dimension

Example

Consider the polynomial matrix $S_v^{(6)}$

$$S_v^{(6)} = \left(\begin{bmatrix} 1. \\ -3. \\ 2. \\ -3. \\ 2. \\ 0. \end{bmatrix}, \begin{bmatrix} 0\ 0\ 0 & 0\ 0\ 0 & 0\ 0\ 0 \\ 0\ 0\ 2 & 1\ 0\ 0 & 0\ 0\ 0 \\ 1\ 0\ 2 & 0\ 0\ 0 & 0\ 0\ 0 \\ 0\ 0\ 0 & 1\ 0\ 2 & 0\ 0\ 0 \\ 0\ 0\ 0 & 1\ 0\ 0 & 0\ 0\ 0 \\ 0\ 0\ 0 & 1\ 1\ 1 & 0\ 0\ 0 \end{bmatrix} \right)$$

which corresponds to a manipulator with n=3 degrees of freedom. Following the described procedure we can eliminate the zero-element in V^T and the corresponding row of E_v (row 6). Further, due to the trigonometric relation $x_3^2 + x_6^2 = 1$, we can contract the rows 2 and 4 into one:

$$\left(\begin{bmatrix} -3. \\ -3. \end{bmatrix} , \begin{bmatrix} 0 & 0 & 2 & 1 & 0 & 0 & 0 & 0 & 0 \\ 0 & 0 & 0 & 1 & 0 & 2 & 0 & 0 & 0 \end{bmatrix} \right) = ([-3.], [0 \ 0 \ 0 \ 1 \ 0 \ 0 \ 0 \ 0 \ 0])$$

The obtained row is the same as the row 5 of E_v. Thus, these rows can be united into

$$([-1.], [0 \ 0 \ 0 \ 1 \ 0 \ 0 \ 0 \ 0 \ 0]).$$

Finally, we obtain the matrix corresponding to the minimal form

$$S_v^{(3)} = \left(\begin{bmatrix} 1. \\ -1. \\ 2. \end{bmatrix} , \begin{bmatrix} 0 & 0 & 0 & 0 & 0 & 0 & 0 & 0 & 0 \\ 0 & 0 & 0 & 1 & 0 & 0 & 0 & 0 & 0 \\ 1 & 0 & 2 & 0 & 0 & 0 & 0 & 0 & 0 \end{bmatrix} \right).$$

We see that the process of minimization can be formalized and easily applied on a digital computer.

3.5. Nonlinear Model

The procedure for creating a nonlinear dynamic model was explained in Para. 2.3. where all parameters as well as joint coordinates and velocities were treated as numerical quantities. Accordingly, the elements of dynamic model matrices were also obtained numerically. From this viewpoint such an algorithm belongs to the class of numerical methods. Alternatively, we introduced polynomials and polynomial matrices as numeric-symbolic representatives of robot variables (Para. 3.2). Finally, we introduced in Para. 3.3 the algebra of polynomial matrices. In this section we shall derive the analytical robot model applying the developed theory.

Suppose that all numeric quantities of manipulator parameters $K_i = \left(\tilde{Q}_i, \tilde{R}_i, \tilde{E}_i \right)$ and $\mathcal{D}_i = \left(m_i, J_i \right)$, $i \in N$, are given (see Fig. 2.7). Alternatively, we shall suppose that the numerical values of joint coordinates are unknown. For this reason, we shall treat them as symbolic variables.

The basic method for closed-form dynamic robot model derivation consits of 3 phases. First, we compute the vectors describing the configuration of the robot when all joint coordinates are equal to zero. Then, we calculate the vectors related to the configuration with some given joint coordinates, and finally the elements of dynamic model matrices.

As the zero joint coordinates configuration is independent of the vector of joint coordinates q, any position vector can be treated numerically, just as in the basic method (Fig. 2.7). Accordingly, numerical quantities determining the set of kinematic variables

$$K_i = \left(Q_i^o, \ R_i^o, \ E_i^o \right), \qquad i \in N$$

are regarded as known (for q=0).

Let we consider now the phase of calculation the actual configuration of the mechanism (q≠0). Note that the polynomial matrices

$$S_{e_i^o} = \left(\vec{e}_i^{\,o}, \ [0 \ \cdots \ 0] \right), \qquad S_{q_{ij}^o} = \left(\vec{q}_{ij}^{\,o}, \ [0 \ \cdots \ 0] \right), \qquad j=1,2,3.$$

are determined. The finite rotation formula will be applied in the following way. First, we shall perform the rotation of all links simultaneously about the axis $\vec{e}_1$ by the angle q_1. During rotating the kinematic chain we suppose that the relative disposition of links remains unchanged, although the disposition of the chain with respect to reference frame is altered. Further, let us perform the rotation of all links except for the first about $\vec{e}_2$ for q_2. Then, similarly, we rotate all links except for the first 2 about $\vec{e}_3$ for q_3, etc. (Fig. 1.2). Let us denote the index of joint about which we perform the rotation by I. The polynomial matrix of $\vec{q}_{ij}^{\,I}$ before the rotation about $\vec{e}_I$ will be denoted as $S_{q_{ij}^I}$. Let us introduce the polynomial matrices S_{e_I} and $S_{q_I} = S_{e_I} \times S_{q_{ij}^I}$ corresponding to the configuration of our manipulator before the rotation about $\vec{e}_I$. Now, we can apply the finite rotation formula in analytic form (for i>I)

$$S_{q_{ij}^{(I+1)}} = \left(S_{e_I} \times S_q x_I + S_q x_{n+I} + \left(S_{e_I} \cdot S_{q_{ij}^I} \right) S_{e_I} \right) \bar{\xi}_I + S_{q_{ij}^I} \xi_I \tag{3.5.1}$$

In order to apply the relation (3.5.1) for $I = 1,\ldots,n$, it is necessary to form $S_{e_{I+1}}$. Using (2.2.3), we obtain

$$\vec{e}_{I+1} = \left[\vec{q}_{I1}^{\,(I+1)} \ \ \vec{q}_{I2}^{\,(I+1)} \ \ \vec{q}_{I3}^{\,(I+1)} \right] \tilde{e}_{I+1}$$

or

$$e_{I+1} = \sum_{\ell=1}^{3} \tilde{e}_{I+1}^{(\ell)} \; \vec{q}_{I\ell}^{\,(I+1)} \tag{3.5.2}$$

where $\tilde{e}_{I+1}^{(\ell)}$ represents the ℓ-th component of $\tilde{e}_{I+1}$. As $\tilde{e}_{I+1}$ represents the vector of parameters whose components are numerical quantities, the expression (3.5.2) becomes in numeric-symbolic domain

$$S_{e_{I+1}} = \sum_{\ell=1}^{3} \tilde{e}_{I+1}^{(\ell)} \; S_{q_{I\ell}^{(I+1)}} \tag{3.5.3}$$

Using (3.5.1) and (3.5.3), which are recursive with respect to the subscript I, we easily obtain the polynomial matrices of vectors from Q_i and E_i, $i \in N$. From the collection of kinematic variables, it remains to form the polynomial matrices of $R_i = \left\{ \vec{r}_{ii}, \; \vec{r}_{i,i+1} \right\}$. As

$$\vec{r}_{ii} = Q_i \tilde{r}_{ii} + q_i \vec{e}_i \xi_i,$$

or

$$\vec{r}_{ii} = \begin{bmatrix} \vec{q}_{i1} & \vec{q}_{i2} & \vec{q}_{i3} \end{bmatrix} \tilde{r}_{ii} + q_i \vec{e}_i \xi_i,$$

or, equivalently,

$$\vec{r}_{ii} = \sum_{\ell=1}^{3} \tilde{r}_{ii}^{(\ell)} \; \vec{q}_{i\ell} + q_i \vec{e}_i \xi_i,$$

the numeric-symbolic representative of $\vec{r}_{ii}$ becomes

$$S_{r_{ii}} = \sum_{\ell=1}^{3} \tilde{r}_{ii}^{(\ell)} \; S_{q_{i\ell}} + \xi_i q_i S_{e_i} \tag{3.5.4}$$

with $\tilde{r}_{ii}^{(\ell)}$ being the ℓ-th component of $\tilde{r}_{ii}$, and

$$\xi_i q_i S_{e_i} = \left(\vec{e}_i, \; E_{e_i} + \begin{bmatrix} 0 & \cdots & 0 & \overset{(2n+i)}{\xi_i} & 0 & \cdots & 0 \\ \vdots & & & \vdots & & & \vdots \\ 0 & \cdots & 0 & \xi_i & 0 & \cdots & 0 \end{bmatrix} \right).$$

Analogously, vector $\vec{r}_{i,i+1} = Q_i \tilde{r}_{i,i+1}$, i.e.

$$\vec{r}_{i,i+1} = \sum_{\ell=1}^{3} \tilde{r}_{i,i+1}^{(\ell)} \vec{q}_{i\ell}.$$

can be written in numeric-symbolic domain as

$$S_{r_{i,i+1}} = \sum_{\ell=1}^{3} \tilde{r}_{i,i+1}^{(\ell)} S_{q_{i\ell}}. \qquad (3.5.5)$$

Thus, the position of all links is determined. In the next step, we shall determine the polynomial matrices of elements of dynamic model matrices. In order to achieve such a goal, consider the variables (3.2.1) which are obtained using nonlinear operations in K_i, $i \in N$. Translating these expressions into numeric-symbolic domain, we get

$$S_{\varepsilon_{ij}} = S_{e_i} \times S_{e_j} \qquad (3.5.6)$$

$$S_{<e,\ r>_{ij}} = S_{e_j} \times S_{r_{ij}} \bar{\xi}_j + S_{e_j} \xi_j \qquad (3.5.7)$$

$$S_{P_{ij\mu}} = S_{e_i} \cdot S_{q_{j\mu}} \qquad (3.5.8)$$

where

$$S_{r_{ij}} = S_{r_{jj}} - S_{r_{j,j+1}} + S_{r_{j+1,j+1}} - \cdots + S_{r_{ii}} \qquad (3.5.9)$$
$$(i>j)$$

Naturally, each algebraic operation in (3.5.6) - (3.5.9) includes minimization of dimensions.

Following the algorithm in Fig. 2.7, we obtain directly

$$S_{H_{ik}} = \sum_{j=\max(i,k)}^{n} \left[m_j S_{<e,\ r>_{ji}} \cdot S_{<e,\ r>_{jk}} + \right.$$

$$\left. + \sum_{\mu=1}^{3} S_{P_{ij\mu}} \cdot S_{P_{kj\mu}} J_{j\mu} \bar{\xi}_i \bar{\xi}_k \right], \qquad (i,k \in N) \qquad (3.5.10)$$

Similarly, we obtain

$$S_{c_{k\ell}^i} = \sum_{j=\max(i,k)}^{n} \left\{ m_j S_{<e,\ r>_{ji}} \cdot \left(S_{e_\ell} \times S_{<e,\ r>_{jk}} \right) + \right. \qquad (3.5.11)$$
$$(k>\ell)$$
$$\left. + \frac{1}{2} \sum_{\mu=1}^{3} \left(S_{P_{ij\mu}} S_{\varepsilon_{\ell k}} + S_{P_{kj\mu}} S_{\varepsilon_{i\ell}} + S_{P_{\ell j\mu}} S_{\varepsilon_{ik}} \right) \cdot S_{q_{j\mu}} J_{j\mu} \bar{\xi}_i \bar{\xi}_k \right\} \bar{\xi}_\ell$$

where $i, k, \ell \in N$.

The elements of gravitational vector $h_G(q, \theta)$ become

$$S_{h_i^G} = -\sum_{j=i}^{n} \left(m_j \, S_{<e, \, r>_{ji}} \cdot \vec{g} \right), \tag{3.5.12}$$

where $\vec{g}$ represents the constant gravity vector. Thus, we obtained elements of any dynamic model matrix in analytic form.

3.6. Linearized and Sensitivity Model

In the second chapter of this book, dealing with model linearization, we obtained the matrices of partial derivatives

$$\frac{\partial H(q, \theta)}{\partial q} \, , \quad \frac{\partial C(q, \theta)}{\partial q} \, , \quad \frac{\partial h_G(q, \theta)}{\partial q}$$

without using numerical differentiation which accumulates errors. This is achieved using 3 theorems stated and proved in Para. 2.5. The developed algorithm, summarized in Fig. 2.10, consists of 2 stages. In the first stage we calculate the partial derivatives of the vectors belonging to the set of kinematic variables. The second stage is devoted to the computation of the partial derivatives of dynamic model matrices. We can now map both stages into symbolic domain using the theorem stated in Para. 3.3. Thus, for example, the first stage becomes

$$S_{<e, \, r>_{ij}^{(s)}} = \left(\left(S_{\varepsilon_{sj}} \times S_{r_{ij}} + S_{\varepsilon_{sj}} \right) \bar{n}_{sj} \bar{\xi}_s + \right.$$

$$\left. + \left(S_{e_j} \times \left(S_{e_s} \times S_{r_{i, \overline{(j,s)}}} \right) \bar{\xi}_s + S_{\varepsilon_{js}} \xi_s n_{sj} \right) \bar{n}_{si} \right) \bar{\xi}_j ,$$

$$S_{\varepsilon_{ji}^{(s)}} = \left(S_{\varepsilon_{si}} \times S_{e_j} \bar{n}_{si} + S_{e_i} \times S_{\varepsilon_{sj}} \bar{n}_{sj} \right) \bar{\xi}_s \tag{3.6.1}$$

$$S_{p_{ij\mu}^{(s)}} = S_{q_{j\mu}} \cdot S_{\varepsilon_{si}} \left(\bar{n}_{si} - \bar{n}_{sj} \right) \bar{\xi}_s$$

with superscript (s) denoting the partial derivative with respect to joint coordinate q_s. Similarly, we can map the equations of the second stage. Although simple conceptually, this solution is not computationaly feasible for the following reason. The number of operations between polynomial matrices is now considerably increased (about 2n times) compared to the basic method for nonlinear model construction. This suggests that it might be well to develop algorithms connected directly to polynomial matrices derivation in symbolic domain. Thus, we suppose that we obtained the polynomial matrices of all elements of dynamic model matrices using a digital computer. Let us denote an arbitrary elelent of a dynamic model matrix by h, and the corresponding polynomial matrix by

$$S_h = \left(h^{(m)}, \ E_h \right)$$

where $h^{(m)} \in R^{m \times 1}$, $E_h \in N^{m \times 3n}$. It turns out that the derivation of the linearized model can be reduced to computation of partial derivatives

$$h^{(s)} = \frac{\partial h(q, \ \theta)}{\partial q_s} \ ,$$

or symbolically

$$S_{h^{(s)}} = \frac{\partial}{\partial q_s} \left(h^{(m)}, \ E_h \right) = \left(h^{(m_s)}, \ E_{h^{(s)}} \right) \tag{3.6.2}$$

The polynomial matrix $S_{h^{(s)}}$, i.e. the partial derivative of S_h with respect to q_s, can be calculated using the following

Theorem

Let $S_h = \left(h^{(m)}, \ E_h \right)$ be the polynomial matrix whose k-th row is given by

$$\left(\left[h_k \right], \ \left[\varepsilon_{1k} \ \cdots \ \varepsilon_{3n,k} \right] \right) \tag{3.6.3}$$

with $k \in \{1, \ldots, m\}$. The polynomial matrix corresponding to the variable $\partial h / \partial q_s$ can be obtained replacing the k-th row (3.6.3) by

$$\left(\begin{bmatrix} -h_k \varepsilon_{sk} \\ h_k \varepsilon_{n+s,k} \\ h_k \varepsilon_{2n+s,k} \end{bmatrix}, \ \begin{bmatrix} \cdots \ \varepsilon_{sk}-1 \ \cdots \ \varepsilon_{n+s,k}+1 \ \cdots & 0 & \cdots \\ \cdots \ \varepsilon_{sk}+1 \ \cdots \ \varepsilon_{n+s,k}-1 \ \cdots & 0 & \cdots \\ \cdots \ 0 \ \cdots \ 0 \ \cdots \ \varepsilon_{2n+s,k}-1 & \cdots \end{bmatrix} \right)$$

$$\tag{3.6.4}$$

for any $k=1, \ldots, m$. Thereafter the minimization of dimensions algorithm can be applied.

Proof

Consider the polynomial

$$h = \sum_{k=1}^{m} h_k x_1^{\varepsilon_{1k}} \cdots x_{3n}^{\varepsilon_{3n,k}} \tag{3.6.5}$$

and delineate the arguments depending on q_s:

$$x_s = \cos q_s$$

$$x_{n+s} = \sin q_s$$

$$x_{2n+s} = q_s.$$

Hence

$$\frac{\partial x_s}{\partial q_s} = -x_{n+s}$$

$$\frac{\partial x_{n+s}}{\partial q_s} = x_s \tag{3.6.6}$$

$$\frac{\partial x_{2n+s}}{\partial q_s} = 1.$$

Differentiating (3.6.5) with respect to q_s and taking into account (3.6.6), we obtain

$$\frac{\partial h}{\partial q_s} = -\frac{\partial h}{\partial x_s} x_{n+s} + \frac{\partial h}{\partial x_{n+s}} x_s + \frac{\partial h}{\partial x_{2n+s}} ,$$

It follows now

$$\frac{\partial h}{\partial q_s} = \sum_{k=1}^{m} \left(-h_k \varepsilon_{sk} x_1^{\varepsilon_{1k}} \cdots x_{s-1}^{\varepsilon_{s-1,k}} x_s^{\varepsilon_{sk}-1} x_{s+1}^{\varepsilon_{s+1,k}} \cdots \right.$$

$$\cdots x_{n+s-1}^{\varepsilon_{n+s-1,k}} x_{n+s,k}^{\varepsilon_{n+s,k}+1} x_{n+s+1,k}^{\varepsilon_{n+s+1,k}} \cdots x_{3n,k}^{\varepsilon_{3n,k}}$$

$$+ h_k \, \varepsilon_{n+s,k} \, x_1^{\varepsilon_{1k}} \cdots x_{s-1}^{\varepsilon_{s-1,k}} x_s^{\varepsilon_{sk}+1} x_{s+1}^{\varepsilon_{s+1,k}} \cdots$$

$$\cdots x_{n+s-1,k}^{\varepsilon_{n+s-1,k}} x_{n+s,k}^{\varepsilon_{n+s,k}-1} x_{n+s+1,k}^{\varepsilon_{n+s+1,k}} \cdots x_{3n,k}^{\varepsilon_{3n,k}}$$

$$+h_k \varepsilon_{2n+s,k} x_1^{\varepsilon_{1k}} \cdots x_{2n+s-1,k}^{\varepsilon_{2n+s-1,k}} x_{2n+s,k}^{\varepsilon_{2n+s,k}-1} x_{2n+s+1,k}^{\varepsilon_{2n+s+1,k}} \cdots x_{3n,k}^{\varepsilon_{3n,k}} \Bigg)$$

which proves the theorem.

Note that the practical usage of this theorem is very simple. It follows from (3.6.4) that 3m multiplications and 5m additions/subtractions have to be employed to compute $\partial h / \partial q_s$. On the other hand observe that the term $h_k \varepsilon_{sk}$ in (3.6.4) actually contains floating-point additions only because ε_{sk} belongs to the set $\{0, 1, 2\}$. Since the same holds for the terms $h_k \varepsilon_{n+s,k}$ and $h_k \varepsilon_{2n+s,k}$, we see that it is not necessary to apply either floating-point or fixed-point multiplication for the calculation of partial derivative $\partial h / \partial q_s$. Since the matrix of exponents usually contains many zero elements, the number of additions is much smaller that the maximal one, mentioned above. The number of rows of the obtained polynomial matrix $S_h(s)$ equals 3m. Since it is not a minimum-dimensional matrix, we have to employ the algorithm described in Para. 3.4. (Fig. 3.1), which yields

$$\underline{S}_h^{(M)}(s) = \min_{m'} S_h^{(m')}(s)$$

where M is the minimal dimension, M<3m.

Example

Let us calculate the partial derivative of the polynomial matrix

$$S_h^{(3)} = \left(\begin{bmatrix} 1. \\ 1. \\ 2. \end{bmatrix} , \begin{bmatrix} 0 & 0 & 0 & 0 & 0 & 0 & 0 & 0 & 0 \\ 0 & 0 & 0 & 1 & 0 & 0 & 0 & 0 & 0 \\ 1 & 0 & 2 & 0 & 0 & 0 & 0 & 0 & 0 \end{bmatrix} \right)$$

with respect to q_1. Following directly the above theorem, we obtain

$$S_h^{(2)}{}_{(1)} = \left(\begin{bmatrix} 1. \\ -2. \end{bmatrix} , \begin{bmatrix} 1 & 0 & 0 & 0 & 0 & 0 & 0 & 0 & 0 \\ 0 & 0 & 2 & 1 & 0 & 0 & 0 & 0 & 0 \end{bmatrix} \right)$$

which has obviously the minimal number of rows.

Let us now derive the closed-form sensitivity model in analytic form. As was pointed out in Para. 2.6, the sensitivity model determines the increment of driving torque (force) – vector necessary to compensate the dynamic-parameter variations of a robot segment $(i=s)$

$$\Delta\theta_s = \left\{\Delta m_s,\ \Delta J_{s1},\ \Delta J_{s2},\ \Delta J_{s3}\right\}$$

under the condition that q, $\dot{q}$ and $\ddot{q}$ do not deviate from their nominal values. It is shown that the link dynamic parameters figure linearly as follows

$$\Delta P = \sum_{\mu=0}^{n_d} \left(H_s^\mu(q)\ddot{q} + \dot{q}^T C_s^\mu(q)\dot{q} + h_s^{G\mu}(q) \right)\Delta\theta_s(\mu+1)$$

with H_s^μ, C_s^μ and $h_s^{G\mu}$ determined by expressions summarized in Fig. 2.12. Examining these expressions we see that they are formed by the use of nonlinear combinations between kinematic variables belonging to sets K_i, $i\in N$. Translating these expressions into analytic form, we obtain relations (3.5.6) – (3.5.9). We henceforth obtain the elements of inertial matrix

$$S_{H^o_{sik}} = S_{\langle e,\ r\rangle_{si}} \cdot S_{\langle e,\ r\rangle_{sk}}\ \overline{\eta_{(i,k)},s}$$

$$S_{H^\mu_{sik}} = S_{P_{is\mu}}\ S_{P_{ks\mu}}\ \overline{\xi}_i\overline{\xi}_k\overline{\eta}_{(i,k)},s$$

(3.6.7)

with $i,k\in N$, $\mu=\{1,2,3\}$.

Similarly, we obtain the elements of C-matrices in the form

$$S_{C^{io}_{sk\ell}} = S_{\langle e,\ r\rangle_{si}} \cdot \left(S_{e_\ell} \times S_{\langle e,\ r\rangle_{sk}}\right)\overline{\xi}_\ell\overline{\eta}_{(i,k)},s$$

(3.6.8)

$$S_{C^{i\mu}_{sk\ell}} = \frac{1}{2}\left(S_{P_{is\mu}}S_{\varepsilon_{\ell k}} + S_{P_{ks\mu}}S_{\varepsilon_{i\ell}} + S_{P_{\ell s\mu}}S_{\varepsilon_{ik}}\right)\cdot S_{q_{s\mu}}\ \overline{\xi}_i\overline{\xi}_k\overline{\xi}_\ell\overline{\eta}_{(i,k)},s$$

(3.6.9)

with $i,k,\ell\in N$, $\mu\in\{1,2,3\}$.

Finally, we obtain the elements of gravitational vector in the form

$$S_{h^G_s(i)} = S_{\langle e,\ r\rangle_{si}}\ \vec{g}\overline{n}_{is}.$$

(3.6.10)

Let us note that the sensitivity model matrices satisfy the same properties as the corresponding nonlinear model matrices. Let us systematize them:

1) Polynomial matrices of the sensitivity model are independent of dy-

namic parameters.

2) Matrix $S_{H_{sik}^{\mu}}$ ($\mu=0,\ldots,3$) is symmetric and positive definite.

3) Matrix $S_{C_{sk\ell}^{i\mu}}$ is symmetric with respect to indices k and ℓ, and anti-symmetric with respect to superscript i and subscript k, i.e.

$$S_{C_{sk\ell}^{i\mu}} = -S_{C_{si\ell}^{k\mu}}, \qquad k,i>\ell.$$

3.7. Approximate Models

As pointed out in preceding sections, approximate mathematical models of robot manipulators play an essential role in real-time applications on moderate or even advanced microcomputers. These models contain fewer floating-point multiplications/additions than exact models. The error which we predefine can be introduced explicitly, using a mathematical criterion [28, 29], or impliictly, neglecting some dynamic effect [59 - 60]. In this section we shall propose a simple procedure suitable for obtaining approximate models in analytic form.

Let $S_{H_{ik}}$, $S_{C_{k\ell}^{i}}$ and $S_{h_{i}^{G}}$ be a set of dynamic model matrices in analytic form. Let us denote an element of a dynamic model matrix by h, and the corresponding polynomial matrix as $S_h = \left(h^{(m)}, E_h \right)$ with m being the number of rows. This matrix is equivalent to the polynomial

$$h(x) = \sum_{k=1}^{m} h_k \, x_1^{\varepsilon_{1k}} \cdots x_N^{\varepsilon_{Nk}},$$

where $x_i = \cos q_i$, $x_{n+i} = \sin q_i$, $x_{2n+i} = q_i$ for $i \in N$. Let us now intro-duce the linear mapping from the set of arguments $X = \{x_i: i=1,\ldots,N\}$ into the domain $Z = \{z_i: i=1,\ldots,N\}$ defined by the following relation

$$z_i = a_i x_i, \qquad i=1,\ldots,N \qquad (3.7.1)$$

where the coefficients a_i are given by

$$a_i = 1, \qquad i=1,\ldots,2n \qquad (3.7.2)$$

$$a_i = \sup_t |q_i(t)|, \quad t \in R^+, \quad i = 2n+1, \ldots, 3n \tag{3.7.3}$$

If mechanical design constraints on joint coordinates are given in the form

$$\underline{q}_i < q_i < \bar{q}_i$$

then eq. (3.7.3) becomes

$$a_i = \max\left(|\underline{q}_i|, |\bar{q}_i|\right). \tag{3.7.4}$$

We see that the described mapping actually means the normalization of variables, i.e. it holds

$$|z_i| < 1, \quad i = 1, \ldots, N. \tag{3.7.5}$$

Substituting (3.7.1) into the polynomial $h(x)$, we obtain

$$h(z) = \sum_{k=1}^{m} h_k^z z_1^{\varepsilon_{1k}} \cdots z_N^{\varepsilon_{Nk}} \tag{3.7.6}$$

where $h_k^z = h_k a_1^{\varepsilon_{1k}} \cdots a_N^{\varepsilon_{Nk}} = h_k a_{2n+1}^{\varepsilon_{2n+1}} \cdots a_{3n}^{\varepsilon_{3n}}$. From the condition (3.7.5) it follows that

$$\sup_{z_i \in Z} |h_k^z z_1^{\varepsilon_{1k}} \cdots z_N^{\varepsilon_{Nk}}| < |h_k^z|. \tag{3.7.7}$$
$$(i = 1, \ldots, N)$$

Let us proceed henceforth to the normalization of the coefficients h_k^z. For this purpose, we shall first determine $k \in \{1, \ldots, m\}$ for which $|h_k^z|$ is maximal

$$\bar{h}^z = \max_k |h_k^z| \tag{3.7.8}$$

and then normalize coefficients of the polynomial (3.7.6), as follows

$$h(z) = \bar{h}^z \sum_{k=1}^{m} \chi_k^z z_1^{\varepsilon_{1k}} \cdots z_N^{\varepsilon_{Nk}} \tag{3.7.9}$$

with $\chi_k^z = h_k^z / \bar{h}^z$. Obviously

$$|\chi_k^z| < 1, \quad k \in \{1, \ldots, m\}. \tag{3.7.10}$$

The corresponding polynomial matrix becomes

$$h(z) = \left(\bar{h}^z \begin{bmatrix} x_1^z \\ \vdots \\ x_m^z \end{bmatrix} , \begin{bmatrix} \varepsilon_{11} & \cdots & \varepsilon_{N1} \\ \vdots & & \vdots \\ \varepsilon_{1N} & \cdots & \varepsilon_{NN} \end{bmatrix} \right) \tag{3.7.11}$$

Note that the matrix of exponents corresponds to the Z domain.

Let us now establish a definition of approximate polynomial function. The function $h_\varepsilon(z)$ is said be approximate to the function $h(z)$ with the error less or equal to ε, if the following inequality holds

$$|h(z) - h_\varepsilon(z)| < \bar{h}^z \varepsilon \tag{3.7.12}$$

for $\forall z \in Z$ and $\varepsilon > 0$.

Observe that ε actually represents the maximal relative error.

Now, we can establish the following theorem:

Theorem

The approximate polynomial $h_\varepsilon(z)$ with a relative error not exceeding ε with respect to the function $h(z)$ is of the form

$$h_\varepsilon(z) = \bar{h}^z \sum_{k \in K^c} x^z \, z_1^{\varepsilon_{1k}} \cdots z_N^{\varepsilon_{Nk}} \tag{3.7.13}$$

where K^c represents the complement of the set $K \subset \{1,\ldots,m\}$, defined by the following inequality

$$\sum_{k \in K} |x_k^z| < \varepsilon. \tag{3.7.14}$$

Proof:

Starting from the well-known inequality

$$\left| \sum_{k \in K} x_k^z \right| < \sum_{k \in K} |x_k^z| \tag{3.7.15}$$

and noting that $|z_i| < 1$, $i \in \{1,\ldots,n\}$, we obtain

$$\left| \sum_{k \in K} \chi_k^z \, z_1^{\varepsilon_{1k}} \cdots z_N^{\varepsilon_{Nk}} \right| < \sum_{k \in K} \left| \chi_k^z \right|. \tag{3.7.16}$$

On the other hand, according to $\cdot$(3.7.13) we have

$$\sum_{k \in K} \chi_k^z \, z_1^{\varepsilon_{1k}} \cdots z_N^{\varepsilon_{Nk}} = \Big(h(z) - h_\varepsilon(z) \Big) / \bar{h}^z, \tag{3.7.17}$$

Substituting into (3.7.16), it follows that

$$\left| h(z) - h_\varepsilon(z) \right| < \bar{h}^z \sum_{k \in K} \left| \chi_k^z \right|. \tag{3.7.18}$$

Taking into account (3.7.14), we see that

$$\left| h(z) - h_\varepsilon(z) \right| < \bar{h}^z \varepsilon$$

which is actually the foregoing definition of approximate polynomials.

The relative error ε can be defined in various other ways. Since h is an element of a dynamic model matrix, the relative error can be defined with respect to the maximal one within an observed matrix. Consider, for example, the inertial matrix elements

$$S_{H_{ik}} = \left(V_{H_{ik}}^{(m_{ik})^T}, \, E_{H_{ik}} \right)$$

with $i \in N$ and $k \in N$. After the normalization of coefficients and variables, we obtain

$$S_{H_{ik}} = \left(\bar{H}_{ik}^z \, v_{H_{ik}}^{(m_{ik})^T}, \, E_{H_{ik}}^z \right)$$

where z indicates the Z domain. The constant $\bar{H}^z$ can be defined as

$$\bar{H}^z = \max_{i,k} \left(\bar{H}_{ik}^z \right). \tag{3.7.19}$$

This constant is common to all elements of inertial matrix.

Further, the relative error can be defined with respect to elements of rows of a model matrix. Then, instead of $\bar{H}^z$, we have

$$\bar{H}_i^z = \max_k \left(\bar{H}_{ik}^z \right) \tag{3.7.20}$$

Thus, we neglect coupling effects (in the sense of defined relative error ε) due to the fact that H_{ik} (k≠i) represents the inertial coupling between i-th and k-th degree of freedom.

We see from the foregoing definitions that the described algorithms are very simple and convenient for computer implementation.

Example

Compute the approximation to polynomial matrix

$$
h = \left(
\begin{bmatrix}
22.5 \\
14.1 \\
0.2 \\
-0.1 \\
7.6
\end{bmatrix}
,
\begin{bmatrix}
1 & 0 & 0 & 0 & 0 & 0 & 0 & 0 & 0 \\
0 & 0 & 0 & 1 & 1 & 0 & 0 & 0 & 0 \\
0 & 1 & 0 & 0 & 0 & 0 & 0 & 0 & 1 \\
0 & 0 & 0 & 0 & 1 & 0 & 0 & 0 & 1 \\
1 & 1 & 0 & 0 & 0 & 0 & 0 & 0 & 0
\end{bmatrix}
.\right)
$$

with the relative error 0.5%. The mechanical design constraints are given with respect to joint coordinate q_3 as

$$-0.3 < q_3 < 0.3.$$

First, we have to translate h(x) into Z domain. Following (3.7.2) and (3.7.4), we have

$$a_i = 1 \qquad i=1,\dots,6$$

$$a_i = 0.3 \qquad i=9.$$

Introducing (3.7.1), we obtain

$$
h = \left(
\begin{bmatrix}
22.5 \\
14.1 \\
0.06 \\
-0.03 \\
7.6
\end{bmatrix}
, E\right)
$$

where E represents the initial matrix of exponents. It will not change during the normalization. Following, further, (3.7.8) - (3.7.9), we get

$$h = \left(22.5 \begin{bmatrix} 1. \\ 0.627 \\ 0.0027 \\ -0.0013 \\ 0.338 \end{bmatrix} , E \right).$$

As $0.0027 + |-0.0013| < \varepsilon = 0.005$, we see that we can eliminate rows 3 and 4 from the polynomial matrix h, and obtain the solution of the stated problem as

$$h_{\varepsilon=0.005} = \left(22.5 \begin{bmatrix} 1. \\ 0.627 \\ 0.338 \end{bmatrix} , \begin{bmatrix} 1 & 0 & 0 & 0 & 0 & 0 & 0 & 0 & 0 \\ 0 & 0 & 0 & 1 & 1 & 0 & 0 & 0 & 0 \\ 1 & 1 & 0 & 0 & 0 & 0 & 0 & 0 & 0 \end{bmatrix} \right)$$

Appendix 3.1

We shall establish here a basic theorem regarding polynomial matrices
of model variables.

Theorem

Consider a robot manipulator in the form of simple open kinematic chain.
Let v denote an arbitrary kinematic or dynamic variable. Then the ele-
ments of E_v matrix of the polynomial matrix $S_v^m = \left(v^{(m)}, E_v \right)$ belong to
the set $\{0, 1, 2\}$.

Proof:

In a preceding section, we showed that the kinematic variables can be
expressed in terms of $\text{Rot}\left(q_1, \ldots, q_n\right)$. We observed that the elements of
Rot are also the elements of Cartesian product (3.2.13). Since there
are no equal elements within the components of the product, it follows
that the maximal value of any exponent cannot exceed 1. Thus, the exponents in polynomial matrices of kinematic variables belong to the set
$\{0, 1\}$.

During the construction of a robot dynamic model (see Fig. 2.7) we encounter the variable $\vec{r}_{ij}$ which can be expressed as

$$\vec{r}_{ij} = \vec{r}_{jj} - \vec{r}_{j,j+1} + \vec{r}_{j+1,j+1} - \cdots + \vec{r}_{ii}, \quad (i>j). \qquad (A3.1.1)$$

Substituting (3.2.16) into the foregoing expression, we obtain

$$\vec{r}_{ij} = \sum_{k=j}^{i-1}\left(R_{kk}^{(3^k)} - R_{k,k+1}^{(3^k)} \right)\text{Rot}\left(q_1, \ldots, q_k\right) + \sum_{k=j}^{i} q_k \vec{e}_k \xi_k$$

$$+ R_{ii}^{(3^i)} \text{Rot}\left(q_1, \ldots, q_i\right). \qquad (A3.1.2)$$

Let us now prove that, if k<i, $\text{Rot}\left(q_1, \ldots, q_k\right)$ is a subset (subvector) of

$\text{Rot}\left(q_1, \ldots, q_i\right)$. Letting $i=k+1$, the set $\text{Rot}\left(q_1, \ldots, q_i\right)$ becomes

$$\left(\cos q_{k+1}, \; \sin q_{k+1}, \; 1\right) \times \left(\cos q_k, \; \sin q_k, \; 1\right) \times \cdots$$

$$\cdots \times \left(\cos q_1, \; \sin q_1, \; 1\right) \qquad (A3.1.3)$$

Since the Cartesian product between the neutral element 1 of the first set and all elements of the remaining sets gives exactly the set $\text{Rot}\left(q_1, \ldots, q_k\right)$, the statement is proved for $i=k+1$. Applying mathematical induction one can easily prove it generally for $k<i$. This property of $\text{Rot}\left(q_1, \ldots, q_k\right)$ will be called "the property of inclusion". Using this property, the expression (A3.1.3) becomes

$$\vec{r}_{ij} = R_{ij}^{(3^i)} \text{Rot}\left(q_1, \ldots, q_i\right) + \sum_{k=j}^{i} q_k \vec{e}_k \xi_k. \qquad (A3.1.4)$$

Now, we see that the exponents of this variable belong to the set $\{0, 1\}$.

Let us now consider the variables (3.2.1) obtained by nonlinear algebraic operations between the elements of K_i, $i \in N$. For example, the vector $\vec{\varepsilon}_{ij} = \vec{e}_i \times \vec{e}_j$ $(i>j)$ represents such a variable. Here, $\vec{e}_i$ and $\vec{e}_j$ are given with respect to the reference frame. First, we shall determine $\tilde{\varepsilon}_{ij}$ with respect to local coordinate system Q_{j-1}. Vector $\tilde{e}_j$ is obviously constant in Q_{j-1}, whereas $\tilde{e}_i$ is of the form (see Fig. 3.2)

$$\tilde{e}_i = \tilde{E}_i^{(3^\ell)} \text{Rot}\left(q_j, \ldots, q_i\right), \qquad \ell=i-j \qquad (A3.1.5)$$

Multiplying $\tilde{e}_i$ by the constant vector $\tilde{e}_j$, we obtain $\tilde{\varepsilon}_{ij}$ as

$$\tilde{\varepsilon}_{ij} = \tilde{\mathcal{E}}_{ij}^{(3^\ell)} \text{Rot}\left(q_j, \ldots, q_i\right), \qquad \ell=i-j. \qquad (A3.1.6)$$

where $\tilde{\mathcal{E}}_{ij}^{(3^\ell)}$ represents a constant vector with respect to Q_{j-1}. Since any vector $\vec{a}_{j-1}$, which is constant with respect to Q_{j-1}, can be presented in the form

$$\vec{a}_{j-1} = A_{j-1}^{(3^{j-1})} \text{Rot}\left(q_1, \ldots, q_{j-1}\right),$$

then any element of $\tilde{\mathcal{E}}_{ij}^{(3^\ell)}$ can be presented in the same manner. Thus, we can write

$$\vec{\varepsilon}_{ij} = \mathcal{E}_{ij}^{(3^i)} \text{Rot}\left(q_1, \ldots, q_i\right), \qquad i>j. \qquad (A3.1.7)$$

138

In other words, exponents in the polynomial matrix of ε_{ij} belong to the set $\{0, 1\}$.

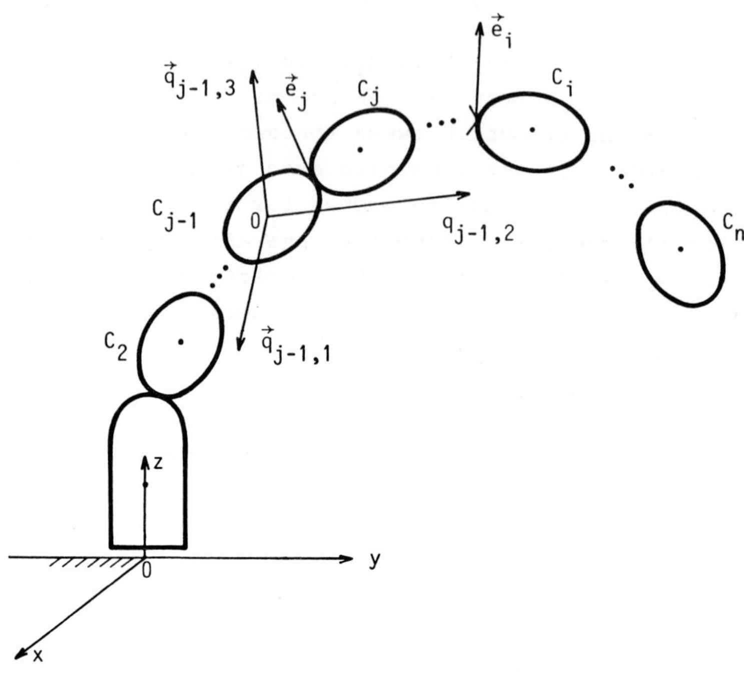

Fig. 3.2. Vectors for determining the polynomial matrix of variable $\vec{\varepsilon}_{ij}$

We shall consider the variable

$$<\vec{e}, \ \vec{r}>_{ij} = \vec{e}_j \times \vec{r}_{ij} \bar{\xi}_j + \vec{e}_j \xi_j, \qquad i>j,$$

in a similar way. Substituting $\vec{r}_{ij}$ given by (A3.1.1) into the foregoing expression, we obtain

$$<\vec{e}, \ \vec{r}>_{ij} = \sum_{k=j}^{i-1} \left(\vec{e}_j \times \vec{r}_{kk} - \vec{e}_j \times \vec{r}_{k,k+1} \right) \bar{\xi}_j + \vec{e}_j \times \vec{r}_{ii} \bar{\xi}_j + \vec{e}_j \xi_j \qquad (A3.1.8)$$

Note that $\vec{r}_{kk}$ and $\vec{r}_{k,k+1}$ are constant with respect to Q_k. Vectors $\vec{e}_j \times \vec{r}_{kk}$ and $\vec{e}_j \times \vec{r}_{k,k+1}$ can be expressed, using $\text{Rot}\left(q_1, \ldots, q_k \right)$, in exactly the same way as $\vec{\varepsilon}_{ij}$. Having in mind the property of inclusion of the Rot – function, we see that the vector $<\vec{e}, \ \vec{r}>_{ij}$ can be expressed in terms of $\text{Rot}\left(q_1, \ldots, q_i \right)$. Thus, the exponents corresponding to the polynomial matrix of $<\vec{e}, \ \vec{r}>_{ij}$ belong to the set $\{0, 1\}$. The same holds for the vector $\vec{e}_\ell \times <\vec{e}, \ \vec{r}>_{ij}$, $\ell<j<i$.

Finally, we consider the variable $p_{ij\mu} = \vec{e}_i \cdot \vec{q}_{j\mu}$. As $\vec{e}_i$ and $\vec{q}_{j\mu}$ represent constant vectors with respect to Q_i and Q_j, respectively, one can proceed as in the case of $\vec{\varepsilon}_{ij}$. Accordingly, the exponents of corresponding polynomial matrices belong to the set $\{0, 1\}$.

Let us return to the expressions for dynamic model matrices, which are summarized in Fig. 2.7. For example, element (i, k) of the matrix $H(q, \theta)$ contains the product $<\vec{e}, \vec{r}>_{ji} \cdot <\vec{e}, \vec{r}>_{jk}$, $j>i>k$. Both of the vectors can be expressed in terms of $\mathrm{Rot}\left(q_1, \ldots, q_j\right)$. Since the corresponding exponents belong to $\{0, 1\}$, it follows that the exponents in the polynomial matrix of their dot product cannot exceed 2. A similar conslusion can be derived for other elements of dynamic model matrices. Thus, the exponents of any dynamic or kinematic variable can only be 0, 1 or 2.

Chapter 4
Model Optimization and Real-Time Program-Code Generation

4.1. Introduction

In the two previous chapters we developed the theoretical background needed for computer-aided construction of robot analytical models. It was shown that in the field of analytical modelling one encounters complex trigonometrical functions which can be mapped into multivariable polynomials. We saw that any of these polynomials consists of a sum of monomials with powers belonging to the set $\{0, 1\}$. Now, we can state one of the most important questions from the practical point of view: how to compute the polynomials with a minimal number of floating-point multiplications/additions? The first part of this chapter will be devoted to this problem. In the second part we will develop an algorithm for optimal computation of polynomials. The algorithm is composed of several steps. In such a manner, the determination of total optimum is reduced to a series of successive local optimizations. Although, in a general case, this need not lead to an absolute minimum, it extremely simplifies the problem solution. For this reason, the algorithm turned out to be very convenient for computer implementation.

In Paragraph 4.2 we will also introduce the graphical schemes (graphs) which illustrate the algorithms of local optimizations. In essence, these graphs represent the optimal sequences of floating-point multiplications/additions. Matrix representatives of graphs will be defined, too. Further, we may state the problem of polynomial calculation starting from the matrix representatives. The solution will be incorporated in a theorem which will show that the computing procedure reduces to a sequence of recursive relations. Each relation contains at most one multiplication and one or more additions. We will see that the polynomials might be calculated by computer, if these relations were mapped into program statements. It will turn out that the problem of mapping can be efficiently solved by computer itself. The computer program accomplishing such a task will be referred to as an "expert-program". The importance of an expert-program lies in a fully automated construction of robot mathematical model.

The obtained (computer-generated) program can thereafter be compiled and used for real-time applications, design purposes, simulations and control synthesis. The program language of the output-file need not be the same as the language used for expert-program design. For instance, the expert-program may be written in a high-level program language (like FORTRAN, PASCAL, etc.), while the output-file may be in ASSEMBLER of a desired microcomputer.

At the end of this chapter we will present several examples illustrating the function of some expert-programs.

4.2. Optimal Computation of Polynomial Matrices

Consider the polynomial matrix $h = \left(h^{(m)^T}, \ E_h \right)$ which corresponds to

$$h(x) = \sum_{k=1}^{m} h_k \ x_1^{\varepsilon_{1k}} \ \cdots \ x_N^{\varepsilon_{Nk}} \tag{4.2.1}$$

According to the theorem formulated and proved in Appendix 3.1, for any robot manipulator it holds

$$\varepsilon_{ik} \in \{0, \ 1, \ 2\}, \qquad i \in \{1, \ldots, N\}, \qquad k \in \{1, \ldots, m\}.$$

Let us introduce now the mapping between the domain of arguments $X = \{x_i : \ i=1, \ldots, N\}$ and the space $Y = \{y_j : \ j=1, \ldots, 2N\}$ such that

$$y_{2i-1} = x_i, \qquad y_{2i} = x_i^2$$

for $i=1, \ldots, N$. Using this nonlinear mapping polynomial (4.2.1) becomes

$$h(y) = \sum_{k=1}^{m} h_k \ y_1^{\sigma_{1k}} \ \cdots \ y_M^{\sigma_{Mk}} \tag{4.2.2}$$

with $M = 2N$. Comparing the latter expression to (4.2.1), we get the σ's as

$$\sigma_{2i-1,k} = \begin{cases} 0 & \varepsilon_{ik} = 0,2 \\ 1 & \varepsilon_{ik} = 1 \end{cases}$$

$$\tag{4.2.3}$$

$$\sigma_{2i,k} = \begin{cases} 0 & \varepsilon_{ik} = 0,1 \\ 1 & \varepsilon_{ik} = 2 \end{cases}$$

for i=1,...,N and k=1,...,m. Notice that σ's correspond to the set of binary numbers

$$\sigma_{ik} \in \{0, 1\} \qquad (4.2.4)$$

for i=1,...,M and k=1,...,m. Obviously, the matrix of exponents of polynomial (4.2.2):

$$E_h^y = \begin{bmatrix} \sigma_{11} & \cdots & \sigma_{M1} \\ \vdots & & \vdots \\ \sigma_{1m} & \cdots & \sigma_{Mm} \end{bmatrix} \qquad (4.2.5)$$

represents a binary matrix, i.e. the matrix with elements being 0's or 1's. This is a very important property from the practical point of view, because it is possible to store any element of matrix (4.2.5) in only one-bit memory cell. Thus, the whole matrix (4.2.5) can be stored in the memory of

$$m(r+M) = m(r+6n) \quad \text{bits}$$

where r represents the number of bits occupied by a real number, and n the number of manipulator degrees of freedom.

Let us consider now the speed of computation of a polynomial h(x). We see from (4.2.1) that the number of floating-point multiplications equals

$$n_M = \sum_{i=1}^{N} \sum_{k=1}^{m} \varepsilon_{ik} \qquad (4.2.6)$$

while the number of floating-point additions equals $n_A = m$. As $\varepsilon_{ik} \in \{0, 1, 2\}$, it follows that

$$0 < n_M < 2mN = 6mn. \qquad (4.2.7)$$

Hence, we see that in the worst case when $\varepsilon_{ik} = 2$ for i=1,...,N and k=1,...,m the number of floating-point multiplications is equal to 6nm. As the number of multiplications dominantly influences the computational time, we must state the problem of optimal calculation of a polynomial h(x) with a minimal number of multiplications. For this purpose we have to develop an algorithm by which the calculation of polynomial (4.2.1) would be reorganized so as to minimize n_M.

We denote by

$$\varepsilon_k = \left[\varepsilon_{1k} \cdots \varepsilon_{Nk} \right] \qquad (4.2.8)$$

the k-th row of matrix of exponents E_h. On the other hand, let us introduce the ordered set of N numbers

$$\beta_i = \left(b_{1i}, \ldots, b_{Ni} \right) \qquad (4.2.9)$$

such that

$$b_{1i} 3^{N-1} + b_{2i} 3^{N-2} + \cdots + b_{Mi} 3^{0} = i$$

and $b_{ji} \in \{0, 1, 2\}$ for any $j \in \{1, \ldots, N\}$.

It will be shown in the text to follow that the quantity

$$dg \ \beta_i = \sum_{k=1}^{N} b_{ki} \qquad (4.2.10)$$

is also important for the optimization task. The symbol dg will be referred to as "degree of".

Let us define now the relation of inclusion ($\mathcal{L}$) of β_i with respect to ε_k, $k \in \{1, \ldots, m\}$ as $\beta_i \ \mathcal{L} \ \varepsilon_k$ such that

$$\beta_i \ \mathcal{L} \ \varepsilon_k \text{ if and only if } b_{ji} \leqslant \varepsilon_{jk} \text{ for any } j \in \{1, \ldots, N\}. \qquad (4.2.11)$$

We will also make use of the number of rows ε_k (4.2.8) which satisfy the condition $\beta_i \ \mathcal{L} \ \varepsilon_k$. Here the subscript i is given, and $k \in \{1, \ldots, m\}$. This number will be denoted as L_i. Alternatively, L_i may be defined as

$$L_i = \sum_{k=1}^{m} I_{ik}, \qquad (4.2.12)$$

with

$$I_{ik} = \begin{cases} 1 & \beta_i \ \mathcal{L} \ \varepsilon_k \\ 0 & \beta_i \ \bar{\mathcal{L}} \ \varepsilon_k \end{cases} \qquad (4.2.13)$$

where $\bar{\mathcal{L}}$ denotes the relation which is opposite to $\mathcal{L}$.

Let us return to the computation optimization of polynomial (4.2.1).

144

For an arbitrary β_i, $i \in \{1,\ldots,3^N-1\}$ let us determine L_i and dg β_i according to the latter definitions. Then we select that β_i for which we obtain the maximal product (dg $\beta_i)L_i'$, i.e.

$$L_I' \ dg \ \beta_I = \max_i \left(L_i' \ dg \ \beta_i \right) \tag{4.2.14}$$

where

$$L_i' = \begin{cases} L_i - 1 & \text{for} \quad L_i > 1 \\ \\ 0 & \text{for} \quad L_i = 0 \end{cases}$$

Now, we have to extract those rows ε_k which satisfy the condition $\beta_I \pounds \varepsilon_k$. We will denote the corresponding set of subscripts k as

$$\kappa_I = \left\{ k_1^I, \ldots, k_{L_I}^I \right\} \tag{4.2.15}$$

The polynomial $h(x)$ (4.2.1) can now be divided into two ports as follows

$$h(x) = h_{11}(x) + h_{12}(x) \tag{4.2.16}$$

where

$$h_{11}(x) = \sum_{k \in \kappa_I} h_k \ x_1^{\varepsilon_{1k}} \cdots x_N^{\varepsilon_{Nk}}$$

$$\tag{4.2.17}$$

$$h_{12}(x) = \sum_{k \in \kappa_I^C} h_k \ x_1^{\varepsilon_{1k}} \cdots x_N^{\varepsilon_{Nk}}$$

with κ_I^C being the complement of the set κ_I with respect to $\{1,\ldots,m\}$. The polynomial $h_{11}(x)$ can be presented as

$$h_{11}(x) = M_{11}(x) \sum_{k \in \kappa_I} h_k \ x_1^{\varepsilon_{1k} - \beta_{1I}} \cdots x_N^{\varepsilon_{Nk} - \beta_{Nk}} \tag{4.2.18}$$

where

$$M_{11}(x) = x_1^{\beta_{1I}} \cdots x_N^{\beta_{NI}} . \tag{4.2.19}$$

The latter expressions are the consequence of the presumption $\beta_I \pounds \varepsilon_k$ for any $k \in \kappa_I$. Notice that the exponents of the monomials under the sum on the right hand side of (4.2.18) are positive, i.e.

$$\varepsilon_{jk} - \beta_{jk} > 0 \quad \text{for any} \quad j \in \{1,\ldots,N\} \quad \text{and} \quad k \in \kappa_I .$$

We see that the number of multiplications in $M_{11}(x)$ equals

$$\sum_{i=1}^{N} \beta_{iI} = dg\ \beta_{I}.$$

As the monomial $M_{11}(x)$ represents the common term for L_I addends of sum (4.2.18), we obtain that the number of multiplications is reduced by $(L_I - 1)dg\ \beta_i = L_I'\ dg\ \beta_I$. From (4.2.14) we see that this product is the maximal one. This implies that this algorithm is optimal in the sense of minimization of the number of floating-point multiplications. After that, n_M is equal to

$$n_M = n_{M(max)} - L_1'\ dg\ \beta_1, \tag{4.2.20}$$

where $n_{M(max)}$ is given by (4.2.6), and $L_1' \equiv L_I'$ and $dg\ \beta_1 \equiv dg\ \beta_I$.

The second step in the optimization starts with polynomials

$$h_{11}'(x) = \sum_{k \in \kappa_I} h_k\ x_1^{\varepsilon_{1k}'} \cdots x_M^{\varepsilon_{Mk}'} \tag{4.2.21}$$

with $\varepsilon_{jk}' = \varepsilon_{jk} - \beta_{jk}$ $(j=1,\ldots,M)$, and

$$h_{12}(x) = \sum_{k \in \kappa_I^c} h_k\ x_1^{\varepsilon_{1k}} \cdots x_N^{\varepsilon_{Nk}}.$$

For both polynomials we should evaluate β_I, $dg\ \beta_I$, L_I and κ_I according to (4.2.10) – (4.2.15). Thus we obtain

$$\beta_{11},\ dg\ \beta_{11},\ L_{11},\ \kappa_{11} \quad \text{for} \quad h_{11}'(x),$$

$$\beta_{12},\ dg\ \beta_{12},\ L_{12},\ \kappa_{12} \quad \text{for} \quad h_{12}(x).$$

Further, the polynomials should be decomposed into subpolynomials in the way described by sets κ_{11} and κ_{12}, as in (4.2.16) and (4.2.17). Then we get

$$h_{11}'(x) = h_{111}(x) + h_{112}(x)$$

$$\tag{4.2.22}$$

$$h_{12}(x) = h_{121}(x) + h_{122}(x)$$

where both subpolynomials $h_{111}(x)$ and $h_{121}(x)$ can be presented as (4.2.18)

$$h_{111}(x) = M_{111}(x) h'_{111}(x)$$

$$h_{121}(x) = M_{121}(x) h'_{121}(x) \tag{4.2.23}$$

with M_{111} and M_{121} being monomials similar to (4.2.19), and h'_{111} and h'_{121} the corresponding sums.

The letter decomposition ends the second step. Thereafter n_M reduces to

$$n_M = n_{M(max)} - L'_1 \, dg \, \beta_1 - L'_{11} \, dg \, \beta_{11} - L'_{12} \, dg \, \beta_{12}, \tag{4.2.24}$$

and the polynomial $h(x)$ to

$$h(x) = M_{11}\left(M_{111} h'_{111} + h_{112}\right) + M_{121} h'_{121} + h_{122}. \tag{4.2.25}$$

The third step involves the decomposition of polynomials h'_{111}, h_{112}, h'_{121} and h_{122}. The procedure repeats until the polynomials reduce to constants. Thereby $h(x)$ obtains the form which depends on

1) monomials (of the type $M(x) = x_1^{\beta_1} \cdots x_M^{\beta_M}$), and

2) constants h_k, $k=1,\ldots,m$.

This form can be symbolically represented by a nonoriented graph Γ shown in Fig. 4.1. The monomials and constants are assigned to graph nodes. Each floating-point multiplication is associated to a branch. So, if two nodes are connected by a graph branch, the corresponding monomials should be multiplied. If more branches are joined in a node, then the node-monomial is to be multiplied by the sum of polynomials

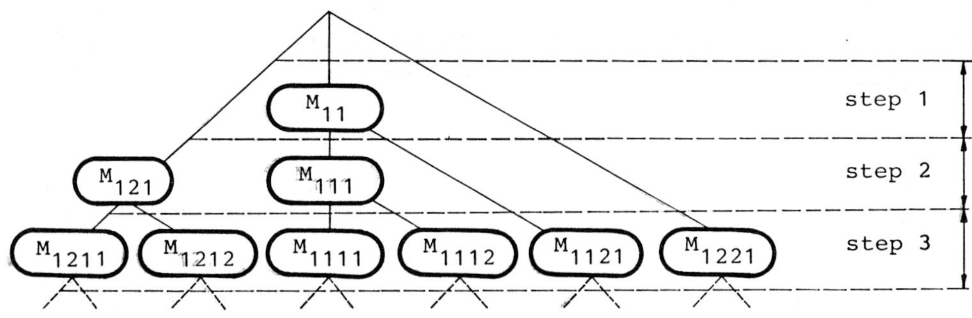

Fig. 4.1. Graphical representation of the polynomial $h(x)$

corresponding to the branches. Accordingly, two basic graph elements
are

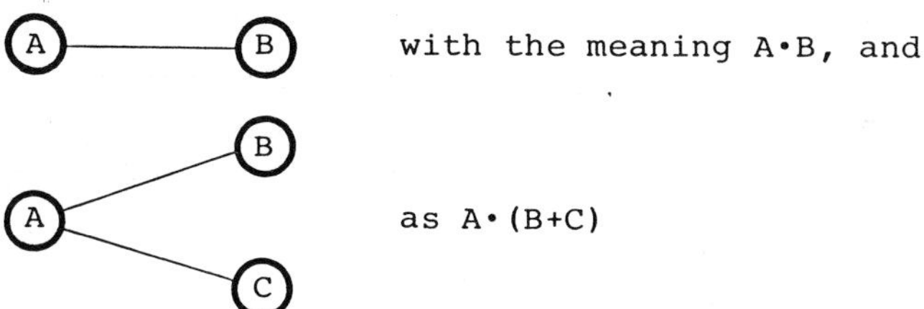

with the meaning A·B, and

as A·(B+C)

Also, if neither a monomial nor a constant is assigned to a node, it
will be assumed that 1 is associated to this node. For example, the
graph

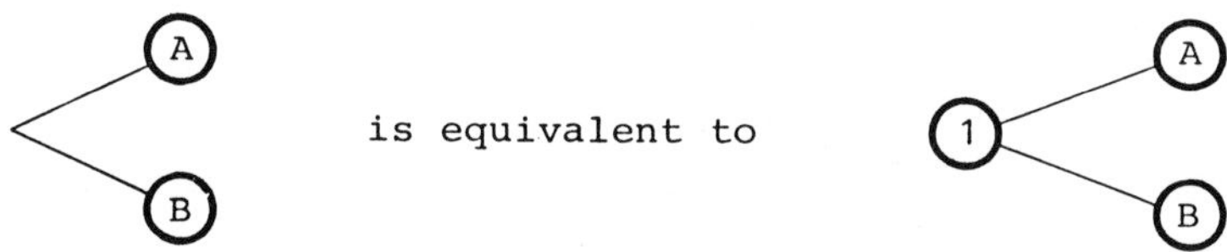

is equivalent to

which corresponds to 1(A+B) = A + B.

For the purpose of computer implementation of the described algorithm,
it is convenient to introduce some matrix representatives of graphs. It
allows treating graphs in a numeric fashion. First, notice that the
node-monomials may be considered as subgraphs. For example, the monomial $M(x) = h_1 x_1 x_3$ can be presented as subgraph

Such graphs will be referred to as "elementary "graphs. To any elementary graph Γ we can associate 2 matrices $E = [e_{ij}]$ and $V = [v_{ij}]$, where
e_{ij} represents the subscript of the variable assigned to the node, and
v_{ij} - the number of branches originating from this node. Denoting the
arguments and constants by X_i, where

$$
X_i = \begin{cases} x_i & 1 \leqslant i \leqslant N \\[2mm] h_{i-N} & N < i \leqslant N + m. \end{cases} \tag{4.2.26}
$$

any variable assigned to a node is uniquely determined by the subscript
i. Forming of matrices E and V for a graph will be demonstrated by a

simple example.

Consider the graph shown in Fig. 4.2. The numbers in nodes are sub-scripts of variables X_i. Here we have N = 9 and m = 7. The nodes can

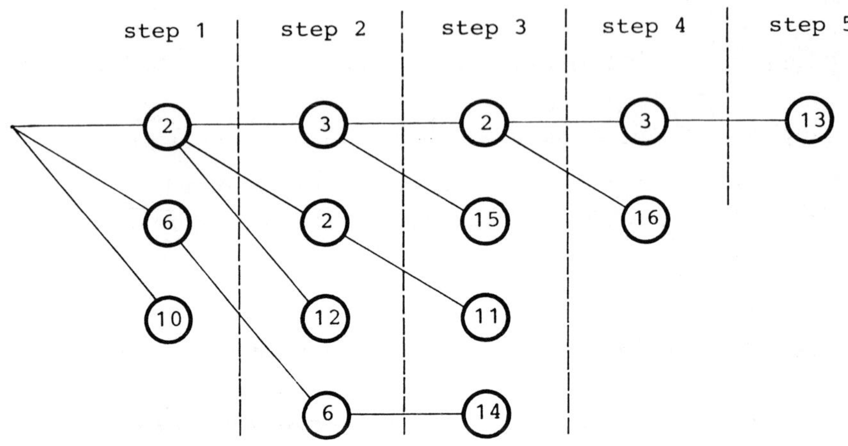

Fig. 4.2. Elementary graph

be mapped in to the E matrix:

$$
E = \begin{bmatrix}
2 & 3 & 2 & 3 & 13 \\
6 & 2 & 15 & 16 & 0 \\
10 & 12 & 11 & 0 & 0 \\
0 & 6 & 14 & 0 & 0
\end{bmatrix} ,
$$

where the number of column corresponds to the number of step. Rows correspond to "parallel" nodes. The same meaning of rows and columns is accepted for V matrix. The elements of V matrix represent the numbers of branches originating from corresponding nodes. Thus we get

$$
V = \begin{bmatrix}
3 & 2 & 2 & 1 & 0 \\
1 & 1 & 0 & 0 & 0 \\
0 & 0 & 0 & 0 & 0 \\
0 & 1 & 0 & 0 & 0
\end{bmatrix}
$$

This example shows how to form E and V matrices for a graph Γ. These matrices uniquely define the optimal sequence of operations for a polynomial computation.

Finally, we can draw the following conclusions:

1) The optimization algorithm is simple and convenient for computer implementation.

2) The optimization is realized step-by-step. Therefore, the resulting number of multiplication is not always absolutely minimal, but nearly-minimal.

3) In some optimization steps more optima might apear, i.e. more subscripts for which $\max_i\left(L_i' \, dg \, \beta_i\right)$ is the same. We might for any of these sequences apply the optimization and finally find the absolute optimum. But this makes the optimization more complicated.

Example

Consider the polynomial matrix

$$
h(x) = \left(
\begin{bmatrix} h_1 \\ h_2 \\ h_3 \\ h_4 \\ h_5 \end{bmatrix}
,
\begin{bmatrix}
1 & 0 & 1 & 0 & 0 & 0 & 0 & 0 & 1 \\
0 & 0 & 1 & 0 & 0 & 0 & 0 & 1 & 1 \\
0 & 0 & 2 & 0 & 2 & 1 & 0 & 0 & 1 \\
0 & 0 & 0 & 0 & 1 & 0 & 0 & 1 & 0 \\
0 & 0 & 2 & 1 & 0 & 0 & 0 & 0 & 1
\end{bmatrix}
\right).
$$

According to (4.2.6) the maximum of n_M equals

$$
n_{M(max)} = 18.
$$

Let us now apply the optimization algorithm clearly indicating the operations within each step.

1. Step

Following the expressions (4.2.9) - (4.2.13) one obtains the table given below:

i	Sequence β_i (4.2.9)	dg β_i (4.2.10)	L_i (4.2.12) - (4.2.13)	L'_i dg β_i
1	0 0 0 0 0 0 0 0 1	1	4	3
3	0 0 0 0 0 0 0 1 0	1	2	1
81	0 0 0 0 1 0 0 0 0	1	2	1
729	0 0 1 0 0 0 0 0 0	1	4	3
730	0 0 1 0 0 0 0 0 1	2	4	6

In this table we indicated only those sequences β_i for which $L'_i > 1$.
Otherwise L'_i dg $\beta_i = 0$. The product L'_i dg β_i is maximal for $\beta_I =$
$= [0\ 0\ 1\quad 0\ 0\ 0\quad 0\ 0\ 1]$, with $I = 730$. The condition $\beta_I \pounds \epsilon_k$ is satisfied for $k = 1,2,3$ and 5, i.e.

$$\kappa_I = \{1,\ 2,\ 3,\ 5\}.$$

The polynomial $h(x)$ now becomes

$$h(x) = M_{11}(x)h'_{11}(x) + h_{12}(x)$$

where, according to (4.2.19)

$$M_{11}(x) = x_3 x_9,$$

and following (4.2.17)

$$h_{12}(x) = h_4 x_5 x_8.$$

According to (4.2.18), $h'_{11}(x)$ becomes

$$h'_{11}(x) = \left(\begin{bmatrix} h_1 \\ h_2 \\ h_3 \\ h_5 \end{bmatrix} , \begin{bmatrix} 1 & 0 & 0 & 0 & 0 & 0 & 0 & 0 & 0 \\ 0 & 0 & 0 & 0 & 0 & 0 & 0 & 1 & 0 \\ 0 & 0 & 1 & 0 & 2 & 1 & 0 & 0 & 0 \\ 0 & 0 & 1 & 1 & 0 & 0 & 0 & 0 & 0 \end{bmatrix} \right)$$

The number of multiplications reduces to

$$n_M = n_{M(max)} - L_i' \, dg \, \beta_I = 12.$$

2. Step

Now we optimize the computation of $h_{11}'(x)$. Polynomial $h_{12}(x)$ is reduced to the optimal form in the previous step.

In $h_{11}'(x)$ there is only one row β_i which satisfies $L_i' > 1$,

$$\beta_I = [0 \ 0 \ 1 \quad 0 \ 0 \ 0 \quad 0 \ 0 \ 0]$$

with $I = 729$, $dg \, \beta_I = 1$ and $L_i' = 1$. As

$$\kappa_I = \{3, \, 4\}$$

$h_{11}'(x)$ can be presented as

$$h_{11}'(x) = M_{111}(x) h_{111}'(x) + h_{112}(x)$$

where

$$M_{111}(x) = x_3,$$

and

$$h_{111}'(x) = \left(\begin{bmatrix} h_3 \\ h_5 \end{bmatrix} , \begin{bmatrix} 0 \ 0 \ 0 & 0 \ 2 \ 1 & 0 \ 0 \ 0 \\ 0 \ 0 \ 0 & 1 \ 0 \ 0 & 0 \ 0 \ 0 \end{bmatrix} \right),$$

$$h_{112}(x) = \left(\begin{bmatrix} h_1 \\ h_2 \end{bmatrix} , \begin{bmatrix} 1 \ 0 \ 0 & 0 \ 0 \ 0 & 0 \ 0 \ 0 \\ 0 \ 0 \ 0 & 0 \ 0 \ 0 & 0 \ 1 \ 0 \end{bmatrix} \right).$$

Obviously, these polynomial matrices cannot be further optimized. Their analytical forms are

$$h_{111}'(x) = h_3 x_5^2 x_6 + h_5 x_4$$

$$h_{112}(x) = h_1 x_1 + h_2 x_8.$$

As $L_I' \, dg \, \beta_I = 1$, n_M is reduced by 1, i.e.

$$n_M = 11.$$

Thereby the optimization is brought to the end. The polynomial $h(x)$ is obtained in the form

$$h(x) = x_3 x_9 \left(x_3 \left(h_3 x_5^2 x_6 + h_5 x_4 \right) + h_1 x_1 + h_2 x_8 \right) + h_4 x_5 x_8$$

which is graphically presented in Fig. 4.3. We see that n_M really equals 11. Accordingly, n_M is reduced from 18 to 11, i.e. by 40%.

Let us now calculate matrix representatives of the graph in Fig. 4.3.

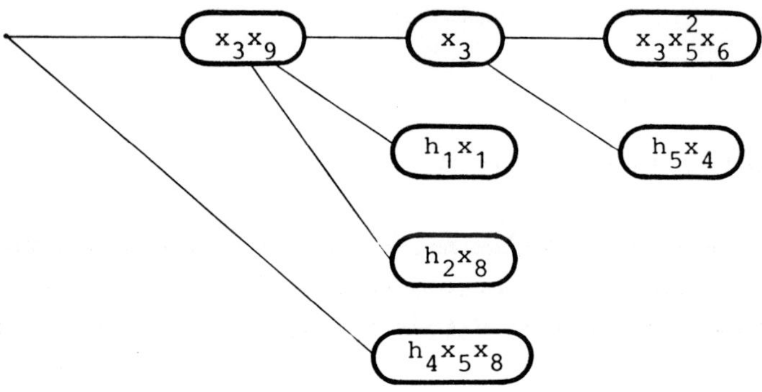

Fig. 4.3. h(x) graph

First, we form the elementary graph (Fig. 4.4) equivalent to the graph in Fig. 4.3. Then, we can directly write E and V matrices

$$E = \begin{bmatrix} 3 & 9 & 3 & 5 & 5 & 6 & 12 \\ 5 & 8 & 1 & 4 & 14 & 0 & 0 \\ 0 & 0 & 8 & 10 & 0 & 0 & 0 \\ 0 & 0 & 13 & 11 & 0 & 0 & 0 \end{bmatrix}$$

$$V = \begin{bmatrix} 1 & 3 & 2 & 1 & 1 & 1 & 0 \\ 1 & 1 & 1 & 1 & 0 & 0 & 0 \\ 0 & 0 & 1 & 0 & 0 & 0 & 0 \\ 0 & 0 & 0 & 0 & 0 & 0 & 0 \end{bmatrix}$$

Notice that any free (end) node of the elementary graph is assigned to a constant number.

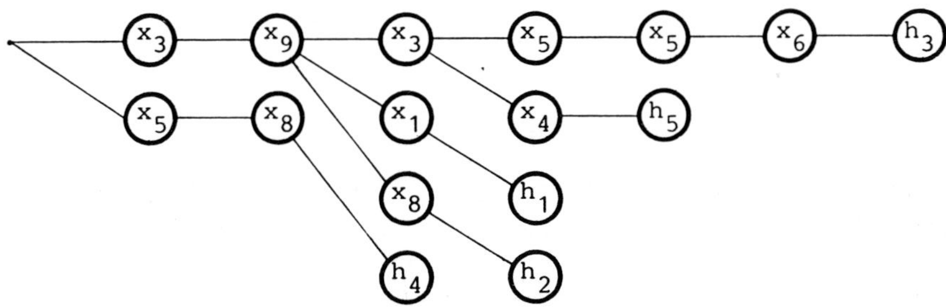

Fig. 4.4. Elementary graph of h(x)

4.3. Real-Time Program-Code Generation

In the previous section we explained how to form the elementary graph
and the corresponding E and V matrices of a polynomial h(x). By using
these matrices it is possible to form a simple algorithm for the polynomial calculation.

Theorem

Let E and V be $n_r \times n_c$ matrix representatives of an elementary graph Γ of
a polynomial h(x). Let us denote by $X_{i,j}$ that variable/constant which
corresponds to the (i, j)-th element of E matrix. We introduce also 2
auxiliary n_r-vectors.

Polynomial h(x) can be computed according to the following algorithm:

1. Step

$$q(i) = X_{i,n_c} \qquad i \in N_n = \left\{1,\ldots,n_r\right\}. \tag{4.3.1}$$

2. Step

$$p(i) = X_{i,n_c-1}\left(q\left(k_{i-1}+1\right)+ \cdots +q\left(k_i\right)\right), \qquad \forall i \in N_n, \tag{4.3.2}$$

where $k_i = \sum_{j=1}^{i} v\left(j, n_c-1\right), \quad k_o = 0.$
$$\vdots$$

(2ℓ)-th step

$$p(i) = X_{i,n_c-2\ell+1}\left(q\left(k_{i-1}+1\right)+ \cdots +q\left(k_i\right)\right), \qquad \forall i \in N_\hbar \qquad (4.3.3)$$

where $k_i = \sum\limits_{j=1}^{i} v\left(j, n_c-2\ell+1\right), \qquad k_o=0.$

(2ℓ+1)-th step

$$q(i) = X_{i,n_c-2\ell}\left(p\left(k_{i-1}+1\right)+ \cdots +p\left(k_i\right)\right), \qquad \forall i \in N_\hbar, \qquad (4.3.4)$$

where $k_i = \sum\limits_{j=1}^{i} v\left(j, n_c-2\ell\right), \qquad k_o=0.$

$\vdots$

(n_c+1)-th step

$$n_c+1 = \begin{cases} \text{even number } - \text{ then } h(x) = \sum\limits_{i=1}^{n_r} p(i) \\[2em] \text{odd number } - \text{ then } h(x) = \sum\limits_{i=1}^{n_r} q(i). \end{cases} \qquad (4.3.5)$$

The proof of this theorem is given in Appendix 4.1.

If should be pointed out that the algorithm described can represent the basis for a computer program which calculates the polynomial h(x). The program which generates the statements according to recursive expressions (4.3.1) - (4.3.5) will be referred to as "expert-program". The output of an expert-program is also a computer program in a high or low-level program language.

From (4.3.1) - (4.3.5) we recognize the basic expression to be generated by an expert-program:

$$p_i = x_j\left(q_{i1}+ \cdots +q_{in}\right)$$

with subscripts i, j and n given. For example, for i=1, j=2 and n=3, let us represent this equation in FORTRAN, ASSEMBLER for microprocessor INTEL 8086 with floating-point coprocessor 8087, and ASSEMBLER for a microcomputer based on microprocessor INTEL 8085.

1) FORTRAN

$$P1 = X2*(QI1+QI2+QI3),$$

2) ASSEMBLER 8086/8087

```
FLD    QI1  ;  QI1→ST
FADD   QI2  ;  ST+QI2→ST
FADD   QI3  ;  ST+QI3→ST
FMUL   X2   ;  ST*X2→ST
FST    P1   ;  ST→P1
```

where ST denotes the first location of STACK - memory, → denotes the data transition. All statements refer to floating-point numbers.

3) ASSEMBLER 8085

```
LXI    H, QI1   ;  QI1→HL
LXI    D, QI2   ;  QI2→DE
CALL   FPADD    ;  QI1+QI2→HL,  exp→A
SHLD   TEMP     ;  HL→TEMP
STA    TEMP+2   ;  A→TEMP+2
LXI    H, TEMP  ;  TEMP→HL
LXI    D, QI3   ;  QI3→DE
CALL   FPADD    ;  QI1+QI2+QI3→HL,  exp→A
SHLD   TEMP     ;  HL→TEMP
STA    TEMP+2   ;  A→TEMP+2
LXI    D, X2    ;  X2→DE
CALL   FPMUL    ;  X2*(QI1+QI2+QI3)→HL,  exp→A
SHLD   P1       ;  HL→P1
STA    P1+2     ;  A→exp→P1
```

where HL, DE and BC are 16-bit registers,

A - 8-bit accumulator

QI1, QI2, QI3 and X2 - memory addresses for floating-point data; the first 16 bits represent the mantissa, and the remaining 8 bits the exponent,

FPADD and FPMUL - subroutines for floating-point addition and multiplication

These examples show that lower-level programming languages require higher complexity of expert-program. Program generation in high-level languages is very simple.

Example

Consider the elementary graph in Fig. 4.4. and the corresponding E and V matrices from the preceding section. The presented theorem generates the following algorithm:

<table>
<tr><td>

$q_1 = x_{12}$

$p_1 = x_6\, q_1$

$q_1 = x_5\, p_1$

$q_2 = x_{14}$

$p_1 = x_5\, q_1$

$p_2 = x_4\, q_2$

$p_3 = x_{10}$

$p_4 = x_{11}$

</td><td>

Contd.

$q_1 = x_3\left(p_1 + p_2\right)$

$q_2 = x_1\, p_2$

$q_3 = x_8\, p_3$

$q_4 = x_{13}$

$p_1 = x_9\left(q_1 + q_2 + q_3\right)$

$p_2 = x_8\, q_4$

$q_1 = x_3\, p_1$

$q_2 = x_5\, p_2$

$H = q_1 + q_2$

</td></tr>
</table>

The FORTRAN code of the algorithm has the form:

```
P1 = X6*X12
Q1 = X5*P1
P1 = X5*Q1
P2 = X4*X14
Q1 = X3*(P1+P2)
Q2 = X1*P2
Q3 = X8*X10
P1 = X9*(Q1+Q2+Q3)
P2 = X8*X13
Q1 = X3*P1
Q2 = X5*P2
H  = Q1+Q2
```

where the statements of the type q=X are omitted for the sake of com-
puter-time minimization.

In ASSEMBLER 8086/8087 this program becomes

```
                                            Contd.
    FLD    X6  ;  X6→ST           FLD    X8   ;  X8→ST
    FMUL   X12 ;  ST*X12→ST       FMUL   X10  ;  ST*X10→ST
    FST    P1  ;  ST→P1           FST    Q3   ;  ST→Q3
    FLD    X5  ;  X5→ST           FLD    Q1   ;  Q1→ST
    FMUL   P1  ;  ST*P1→ST        FADD   Q2   ;  ST+Q2→ST
    FST    Q1  ;  ST→Q1           FADD   Q3   ;  ST+Q3→ST
    FLD    X5  ;  X5→ST           FMUL   X9   ;  ST*X9→ST
    FMUL   Q1  ;  ST*Q1→ST        FST    P1   ;  ST→P1
    FST    P1  ;  ST→P1           FLD    X8   ;  X8→ST
    FLD    X4  ;  X4→ST           FMUL   X13  ;  ST*X13→ST
    FMUL   X14 ;  ST*X14→ST       FST    P2   ;  ST→P2
    FST    P2  ;  ST→P2           FLD    X3   ;  X3→ST
    FLD    P1  ;  P1→ST           FMUL   P1   ;  ST*P1→ST
    FADD   P2  ;  ST+P2→ST        FST    Q1   ;  ST→Q1
    FMUL   X3  ;  X3*ST→ST        FLD    X5   ;  X5→ST
    FST    Q1  ;  ST→Q1           FMUL   P2   ;  ST*P2→ST
    FLD    X1  ;  X1→ST           FST    Q2   ;  ST→Q2
    FMUL   P2  ;  ST*P2→ST        FLD    Q1   ;  Q1→ST
    FST    Q2  ;  ST→Q2           FADD   Q2   ;  ST+Q2→ST
                                  FST    H    ;  ST→H
```

At the end we have to point out that the expert-program can be written
in any programming language.

Appendix 4.1

Here, the theorem stated in Paragraph 4.3 will be proved. For this purpose, consider a general elementary graph Γ shown in Fig. 4.5. The subscripts corresponding to subscripts of E matrix are assigned to nodes. Hence, the node (i, j) corresponds to the element e_{ij}, i.e. the variable X_{ij}. The number of branches v_{ij} originating from any node is indicated, too.

Denote the variables corresponding to nodes $(i,\ n_c)$ as

$$q(i) = X_{i,n_c}, \qquad\qquad i \in N_\hbar .$$

Consider now the node $(1,\ n_c-1)$ from which v_{1,n_c-1} branches originate. These branches end with nodes $(1,\ n_c),\ldots,(v_{1,n_c-1},\ n_c)$. Let us substitute the subgraph containing the node $(1,\ n_c-1)$, all branches originating from this node, and end-nodes, by one node.

Obviously, the polynomial

$$p(1) = X_{1,n_c-1}\Big(q(1) + \cdots + q\big(v_{1,n_c-1}\big)\Big) \tag{P4.1.1}$$

must be assigned to this node. If we apply the same to the nodes $(i,\ n_c-1)$ for $i=2,\ldots,n_r$, we will obtain just the polynomials (4.3.2). Thus, we obtain the graph equivalent to the elementary graph τ containing the nodes $p(1),\ldots,p(n_r)$.

We will prove the other expressions in the theorem in a similar way. In the $(2\ell-1)$-st step, let the nodes $(i,\ n_c-2\ell+1)$, $i \in N_\hbar$, and the branches originating from the nodes, be replaced by equivalent nodes $q(1),\ldots,q(n_r)$. Consider now the node $(1,\ n_c-2\ell+1)$. As in derivation of (P4.1.1), we will obtain that the subgraph consisting of the node, branches starting from the node and going to the right, and nodes in which the branches terminate $q(1),\ldots,q(v_1,\ n_c-2\ell+1)$, can be replaced

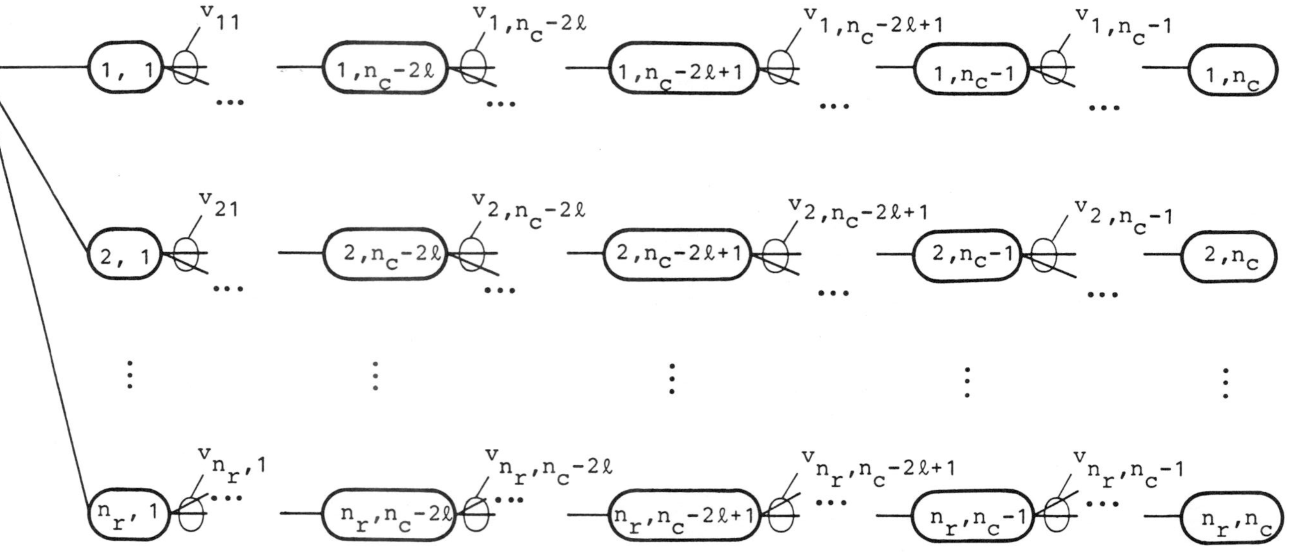

Fig. 4.5. General form of an elementary graph

by a node to which the following polynomial is assigned

$$p(1) = X_{1,n_c-2\ell+1}\Big(q(1) + \cdots + q\big(v_{1,n_c-2\ell+1}\big)\Big). \qquad (P4.1.2)$$

Applying the same procedure to nodes $(i, n_c-2\ell+1)$, we obtain the polynomials (4.3.3) of the (2ℓ)-th step of the theorem.

Expressions (4.3.4) and (4.3.5) can be proved in the same way.

Chapter 5
Examples

5.1. Introduction

All previously presented results will be illustrated in this chapter by
the examples of actual robots implemented in industrial technological
processes. Light robots (UMS-1 and UMS-2, Yugoslav production), as well
as Dependable-Fordath heavy-duty robot, (American production) will be
considered. These robots have been selected on the basis of the structure of their mechanisms. Thus, UMS-2 robot is of cylindrical structure,
UMS-1 of anthropomorphic type (similar to a human arm), while the heavy-
-duty robot is of arthropoid type. Evidently, mechanisms of different
structures have been chosen so as to illustrate the dependence of the
mathematical model complexity on the complexity of structure.

The first example refers to a cylindrical-structure robot ("a cylindrical robot") with one rotational and two translational degrees of freedom. An arthropoid robot with three rotational degrees of freedom, the
axes of the second and third joints being parallel, is considered in
the second example. The third example discusses an anthropomorphic robot with three degrees of freedom and with no parallel axes. A detailed
kinematic scheme with all required geometrical and dynamic parameters
is given for each example. A method for the construction of nonlinear
numeric-symbolic model, described in the preceding chapters, is then
illustrated. This includes the presentation of the outputs of all stages,
such as the stage of determining the initial and actual positions of
the mechanism links, the stage of determining the inertial matrix, C -
matrix, and the stage of determining the gravity vector. The outputs of
these stages will be given through the corresponding polynomial matrices from which polynomial forms, i.e. analytical functions are directly obtainable. The matrices of dynamic model formed in this way are
then used to show that all the properties of the matrices of dynamic
model, such as symmetry, positive definiteness and antisymmetry (Chapter 2), hold.

Each of these three examples is also used to illustrate the method for
forming the linearized numeric-symbolic model, presented in Para. 2.5.

and 3.6. This includes the use of the theorem of partial derivatives of the polynomial matrix, proved in Para. 3.6. This theorem is shown to be very simple and practical for program implementation.

The method for constructing the sensitivity model of robots will also be illustrated by means of these examples. The effect of variations in dynamic parameters (mass, inertial moments) on the driving torques or forces, as described in Para. 2.6 and 3.6, will be considered. It will be shown that sensitivity model computation requires fewer number of numerical operations than the complete manipulator dynamic model calculation.

The procedure for forming approximate robot models will also be presented using the example of arthropoid robot. This procedure was theoretically elaborated in Para. 2.6 and 3.7. In this example it will be shown that it is possible to neglect dynamic effects with respect to dominant effects. For example, Coriolis' effects may often be neglected with respect to gravitational. Estimation of maximum error will also be given in the example, i.e. the effects which can be neglected will be determined for a given permissible error.

The procedure for computational optimization of numeric-symbolic models will be illustrated at the end of each example. All notions introduced in Chapter 4, such as the sequence of optimal calculation, the elementary graph, matrix representation of graph, etc., will also be illustrated. The generation of the program code of model using expert-program will be considered in particular. The exact number of the required floating--point multiplications and additions, which is very important for microcomputer implementation, will then be determined from the generated program codes. A comparison between the numbers of operations required for forming the linearized and the nonlinear model will be provided for one example. All previous theoretical results, i.e. all the modules of the corresponding program package, will thus be illustrated.

5.2. A Cylindrical Robot

Let us consider a manipulator of cylindrical structure, shown in Fig. 5.1. This robot has three basic degrees of freedom powered by electrical motors with permanent magnets. Various devices, such as a gripper with 1 to 3 degrees of freedom, a pneumatic vacuum cup, etc., may be connected to the end of the third link. To systematically il-

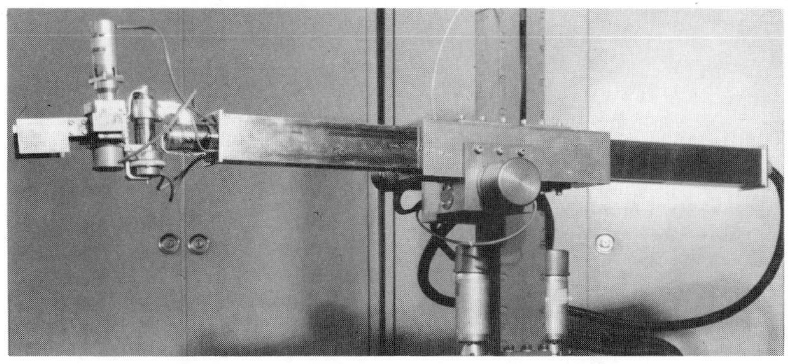

Fig. 5.1. A cylindrical robot

lustrate the algorithms presented in the preceding text, we will first represent the structure of the mechanism of this manipulator by a corresponding model (Fig. 5.2). C_1, C_2 and C_3 denote mechanism links and z_1, z_2 and z_3 joint connections, the first of which is rotational and the second two are translational. Accordingly, this mechanism is of RTT structure. Mechanism links with all parameters are separately represented in Fig. 5.3. As stated in Para. 2.2, each mechanism link may be described by an arranged set of parameters $C_i\left(K_i,\ \mathcal{D}_i\right)$, where $K_i = \left(Q_i,\ \tilde{R}_i,\ \tilde{E}_i\right)$ is a set of kinematic parameters, and $\mathcal{D}_i = \left(m_i,\ J_i\right)$ a set of dynamic parameters. In Fig. 5.3 an internal coordinate system $Q_i = \left(\vec{q}_{i1},\ \vec{q}_{i2},\ \vec{q}_{i3}\right)$ is assigned to each mechanism link. In such coordinate systems, the sets of distance vectors $\tilde{R}_i = \left\{\tilde{r}_{ii},\ \tilde{r}_{i,i+1}\right\}$ and vectors of joint exes $\tilde{E}_i = \left\{\tilde{e}_i\right\}$ are defined. Numerical data for the

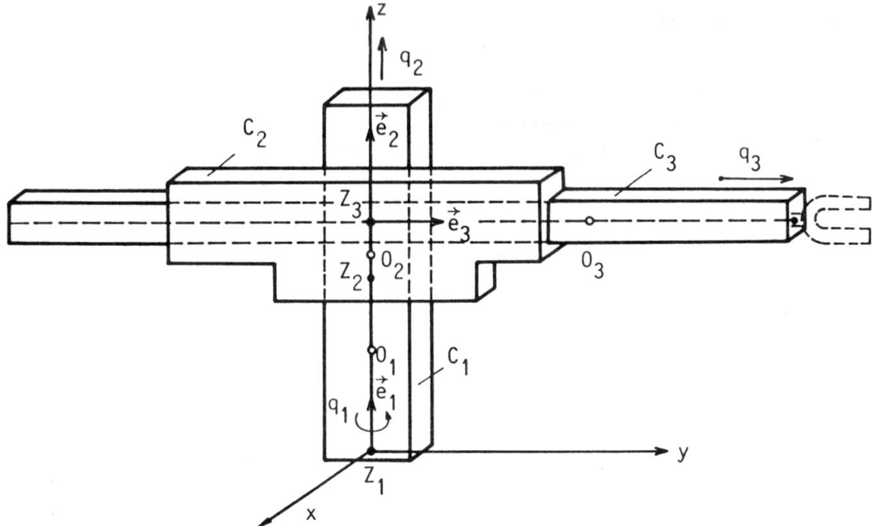

Fig. 5.2. Model of the mechanism of cylindrical robot

actual manipulator are given in Table 5.1. ξ_i denotes the type of the i-th joint; $\xi_i = 0$ for the rotational joint, and $\xi_i = 1$ for the translational joint. It should be noted that all joint connections: $Z_1 = Z_{11}$, $Z_2 = Z_{12} = Z_{21}$ and $Z_3 = Z_{23} = Z_{32}$, i.e. all kinematic pairs $\{C_o, C_1\}$, $\{C_1, C_2\}$ and $\{C_2, C_3\}$ are of the class 5. Pair $\{C_o, C_1\}$, where C_o denotes the support, is of subclass 1, and the remaining two pairs are of subclass 2.

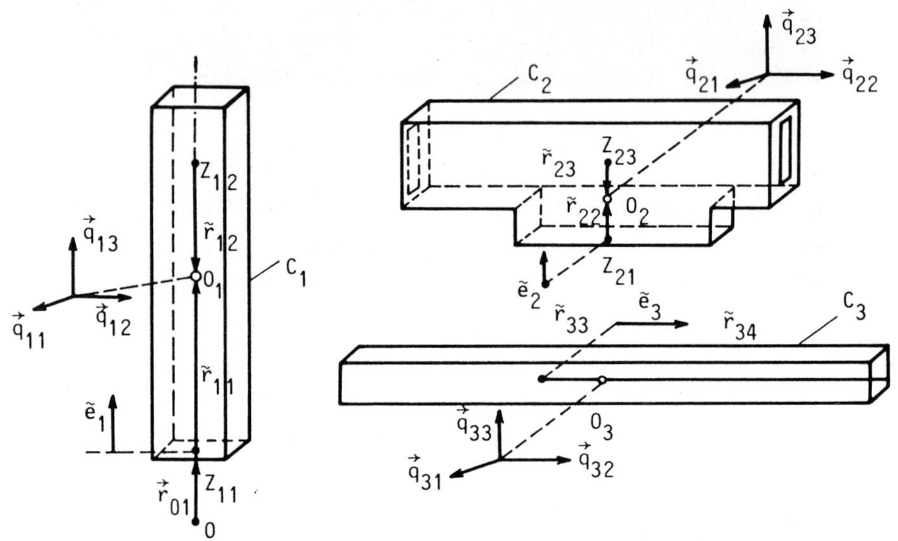

Fig. 5.3. Kinematic parameters of mechanism links
of cylindrical robot

i	$\tilde{r}_{ii}$ [m]	$\tilde{r}_{i,i+1}$ [m]	$\tilde{e}_i$	ξ_i
1	(0., 0., 0.188)	(0., 0., -0.192)	(0., 0., 1.)	0
2	(0., 0., 0.01)	(0., 0., -0.01)	(0., 0., 1.)	1
3	(0., 0.01, 0.)	(0., -0.44, 0.)	(0., 1., 0.)	1

Table 5.1. Kinematic parameters of cylindrical robot

Dynamic parameters of the mechanism are given in Table 5.2. The parameters which are not included in this table are not important for constructing the dynamic model of cylindrical robot. To apply the developed algorithm for the construction of matrices of dynamic robot model in a closed form (see the diagram in Fig. 2.7), it is necessary to specify another 2 input vectors $\vec{r}_{o1}$ and $\vec{e}_1$:

$$\vec{r}_{o1} = (0., 0., 0.),$$

$$\vec{e}_1 = (0., 0., 1.).$$

i	m_i [kg]	J_{i1} [kg m^2]	J_{i2} [kg m^2]	J_{i3} [kg m^2]
1	–	–	–	0.0294
2	7.	–	–	0.0550
3	4.15	–	–	0.3180

Table 5.2. Dynamic parameters of cylindrical robot

Derivation of nonlinear analytical model of cylindrical robot

The first stage of the algorithm includes the computation of the "home" position of the mechanism. On the basis of previous data, the following initial values are obtained

$$Q_i^O = \begin{bmatrix} 1 & 0 & 0 \\ 0 & 1 & 0 \\ 0 & 0 & 1 \end{bmatrix},$$

$$\vec{r}_{ii}^{\,O} = \tilde{r}_{ii}, \qquad \vec{r}_{i,i+1}^{\,O} = \tilde{r}_{i,i+1}, \qquad (i=1,2,3),$$

and

$$\vec{e}_{i+1}^{\,O} = \tilde{e}_{i+1}, \qquad (i=1,2).$$

In the next stage of the algorithm, the position of mechanism links as a function of joint coordinates q_1, q_2 and q_3 is determined. Accordingly, the variables representing the output of this stage are not numerical values, but functions which may be described by polynomial matrices, i.e., by polynomial forms. Para. 3.5 presents the procedure for forming the desired polynomial matrices

$$S_{q_{ij}}, \qquad i \in N, \qquad j=1,2,3,$$

$$S_{r_{ii}}, \quad S_{r_{i,i+1}} \quad \text{and} \quad S_{e_i}.$$

Generating these matrices by computer, one obtains

$$S_{q_{i1}} = \left(\begin{bmatrix} 1.000 & 0.000 & 0.000 \\ 0.000 & 1.000 & 0.000 \end{bmatrix}, \quad \begin{bmatrix} 1 & 0 & 0 & 0 & 0 & 0 \\ 0 & 0 & 0 & 1 & 0 & 0 \end{bmatrix} \right)$$

$$S_{q_{i2}} = \left(\begin{bmatrix} 0.000 & 1.000 & 0.000 \\ -1.000 & 0.000 & 0.000 \end{bmatrix}, \quad \begin{bmatrix} 1 & 0 & 0 & 0 & 0 & 0 \\ 0 & 0 & 0 & 1 & 0 & 0 \end{bmatrix} \right)$$

$$S_{q_{i3}} = \left([0.000 \quad 0.000 \quad 1.000], \quad [0 \ 0 \ 0 \ 0 \ 0 \ 0] \right)$$

for $i = 1,2,3$. It should be noted that the obtained polynomial matrices are of minimal form. Namely, after applying a series of algebraic operations to the polynomial matrices (see Rodrigues' formula in the symbolic domain (3.5.1)), optimization of the dimensions of the obtained forms is performed according to the algorithm presented in Para. 3.4.

The polynomial form of polynomial matrix $S_{q_{i1}} = \left(V_{q_{i1}}^T, \ E_{q_{i1}} \right)$ is directly obtained from

$$V_{q_{i1}}^T = \begin{bmatrix} 1.000 & 0.000 & 0.000 \\ 0.000 & 1.000 & 0.000 \end{bmatrix}$$

$$E_{q_{i1}} = \begin{bmatrix} 1 & 0 & 0 & 0 & 0 & 0 \\ 0 & 0 & 0 & 1 & 0 & 0 \end{bmatrix}.$$

If notation $V_{q_{i1}}^T = \begin{bmatrix} V_1^T \\ V_2^T \end{bmatrix}$ is introduced, we obtain

$$\vec{q}_{i1} = v_1 x_1 + v_2 x_4 = \begin{bmatrix} 1.000 \\ 0.000 \\ 0.000 \end{bmatrix} x_1 + \begin{bmatrix} 0.000 \\ 1.000 \\ 0.000 \end{bmatrix} x_4$$

where, according to (3.4.1)

$$x_1 = \cos q_1$$
$$x_4 = \sin q_1,$$

or

$$\vec{q}_{i1} = \begin{bmatrix} \cos q_1 \\ \sin q_1 \\ 0 \end{bmatrix}.$$

The remaining polynomial matrices are interpreted in a similar way.

The polynomial matrices of kinematic variables $R_i = \left\{ \vec{r}_{ii}, \vec{r}_{i,i+1} \right\}$, $i \in \{1,2,3\}$, are given in Table 5.3. As may be seen, $S_{r_{22}}$ and $S_{r_{33}}$ also depend on the polynomial arguments $x_8 = q_2$ and $x_9 = q_3$, which is a result of the linear motion of the second and third degree of freedom. The polynomial matrices of the vectors of joint axes are shown in Table 5.4. As may be seen these polynomial matrices are extremely simple and only S_{e_3} depends on $x_1 = \cos q_1$ and $x_4 = \sin q_1$, which is a result of the cylindrical structure of the robot. The polynomial matrices of all kinematic variables K_i, $i \in N$ are thus determined.

i	$S_{r_{ii}}$	$S_{r_{i,i+1}}$
1	$([0.\ \ 0.\ \ 0.188],\ [000\ 000\ 000])$	$([0.\ \ 0.\ \ -0.192]\ ,\ [000\ 000\ 000])$
2	$\left(\begin{bmatrix} 0.\ \ 0.\ \ 0.01 \\ 0.\ \ 0.\ \ 1. \end{bmatrix} ,\ \begin{bmatrix} 000\ 000\ 000 \\ 000\ 000\ 010 \end{bmatrix} \right)$	$([0.\ \ 0.\ \ -0.01]\ ,\ [000\ 000\ 000])$
3	$\left(\begin{bmatrix} 0. & 0.01 & 0. \\ -0.01 & 0. & 0. \\ 0. & 1. & 0. \\ -1. & 0. & 0. \end{bmatrix} ,\ \begin{bmatrix} 100\ 000\ 000 \\ 000\ 100\ 000 \\ 100\ 000\ 001 \\ 000\ 100\ 001 \end{bmatrix} \right)$	$\left(\begin{bmatrix} 0. & -0.44 & 0. \\ 0.44 & 0. & 0. \end{bmatrix} ,\ \begin{bmatrix} 100\ 000\ 000 \\ 000\ 100\ 000 \end{bmatrix} \right)$

Table 5.3. Polynomial matrices of distance vectors

The next stage of the algorithm includes the generation of analytical expressions of all elements of the inertial matrix of the mechanism (see Fig. 2.7.) according to (3.5.10). This is how we obtain the polynomial matrices in Table 5.5., where the polynomial matrices of elements H_{ij} for $i \leqslant j$ are presented. The remaining elements are directly obtained from the property of symmetry of the inertial matrix

$$S_{H_{ij}} = S_{H_{ji}}, \qquad i,j \in N,$$

which was proved in Para. 2.4.

i	S_{e_i}
1	$([0.\ 0.\ 1.], \quad [000\ 000\ 000])$
2	$([0.\ 0.\ 1.], \quad [000\ 000\ 000])$
3	$\left(\begin{bmatrix} 0.\ 1.\ 0. \\ -1.\ 0.\ 0. \end{bmatrix}, \begin{bmatrix} 100\ 000\ 000 \\ 000\ 100\ 000 \end{bmatrix}\right)$

Table 5.4. Polynomial matrices of the joint axes unit vectors

i,j	$S_{H_{ij}}$
1,1	$\left(\begin{bmatrix} 0.40281 \\ 0.08300 \\ 4.15000 \end{bmatrix}, \begin{bmatrix} 000\ 000\ 000 \\ 000\ 000\ 001 \\ 000\ 000\ 002 \end{bmatrix}\right)$
1,2	$([0.00000], \quad [000\ 000\ 000])$
1,3	$([0.00000], \quad [000\ 000\ 000])$
2,2	$([11.1500], \quad [000\ 000\ 000])$
2,3	$([0.00000], \quad [000\ 000\ 000])$
3,3	$([4.15000], \quad [000\ 000\ 000])$

Table 5.5. Polynomial matrices of the elements of inertial
matrix of cylindrical robot

The polynomial matrices of the elements of C - matrices of the model
are generated in the next stage according to expression (3.5.11), (see
the diagram in Fig. 2.7). The results are summed in Table 5.6. Element
c_{31}^1 is evidently the same as element c_{13}^1, which results from the pro-

perty of symmetry of matrices $C^i \in R^{n \times n}$, $i \in N$. One may also conclude from the table that the following holds

$$S_{C^1_{13}} = -S_{C^3_{11}}$$

i	k, ℓ	$S_{C^i_{k\ell}}$	
1	1,3	$\left(\begin{bmatrix} 0.04150 \\ 4.15000 \end{bmatrix} \right.$,	$\left. \begin{bmatrix} 000\ 000\ 000 \\ 000\ 000\ 001 \end{bmatrix} \right)$
1	3,1	$\left(\begin{bmatrix} 0.04150 \\ 4.15000 \end{bmatrix} \right.$,	$\left. \begin{bmatrix} 000\ 000\ 000 \\ 000\ 000\ 001 \end{bmatrix} \right)$
3	1,1	$\left(\begin{bmatrix} -0.04150 \\ -4.15000 \end{bmatrix} \right.$,	$\left. \begin{bmatrix} 000\ 000\ 000 \\ 000\ 000\ 001 \end{bmatrix} \right)$

Table 5.6. Polynomial matrices of nonzero elements of
C - matrices of cylindrical robot

This represents the property of antisymmetry, which was proved in Para. 2.4. Now, one concludes that matrices C^i, $i \in N$, have the following form

$$C^1 = \begin{bmatrix} 0 & 0 & C^1_{13} \\ 0 & 0 & 0 \\ C^1_{13} & 0 & 0 \end{bmatrix},$$

$$C^2 = \begin{bmatrix} 0 & 0 & 0 \\ 0 & 0 & 0 \\ 0 & 0 & 0 \end{bmatrix}$$

and

$$C^3 = \begin{bmatrix} -C^1_{13} & 0 & 0 \\ 0 & 0 & 0 \\ 0 & 0 & 0 \end{bmatrix}.$$

The last stage of the algorithm consists of calculating the gravity vector $S_{h_1^G}$, $i \in N$. According to

$$S_{h_1^G} = ([0.00000], [000\ 000\ 000]),$$

$$S_{h_2^G} = ([109.3815], [000\ 000\ 000])$$

and

$$S_{h_3^G} = ([0.00000], [000\ 000\ 000]).$$

The polynomial matrices of all elements of the matrices of the dynamic model for cylindrical robot have thus been determined. For illustration, Table 5.7 presents the analytical forms of all obtained nonzero polynomial matrices. The definition of analytical form (3.2.19) - (3.2.22). has been used for this. Let us note that, according to these definitions, $x_9 = q_3$ is the internal coordinate of the third joint.

	INERTIAL MATRIX	
(i,j)	Polynomial $S_{H_{ij}}$	
$(1,1)$	$0.40281 + 0.08300\ x_9 + 4.15\ x_9^2$	
$(2,2)$	11.1500	
$(3,3)$	4.1500	
	C - MATRIX	
9	(k,ℓ)	Polynomial $S_{C_{k\ell}^i}$
1	$1,3$	$0.04150 + 4.15000\ x_9$
	GRAVITY VECTOR	
i	Polynomial $S_{h_i^G}$	
2	109.38151	

Table 5.7. Polynomial forms corresponding to nonzero elements of dynamic model matrices for cylindrical robot

Construction of linearized model of
cylindrical robot

The algorithm for forming the linearized model of a robot in closed
form (2.5.1) was presented in Para. 2.5 and summed in Figs. 2.9 and
2.10. Two procedures for the construction of the corresponding numeric-
-symbolic model were described in Para. 3.6. The first of them is based
on the algorithm given in Para. 2.5, and the second on the theorem of
differentiation of polynomial matrices. These algorithms are equivalent
and easily implementable on digital computers. The outputs of these algorithms are the following polynomial matrices

$$^{S}H_{ij}(s)' \qquad ^{S}C_{jk}^{i}(s) \qquad \text{and} \qquad ^{S}h_{i}^{G}(s)$$

for $i,j,k,s \in N$. As stated in Para. 3.6, (s) stands instead of the oper-
ator $\frac{\partial}{\partial q_s}$. Table 5.8 shows all nonzero polynomial matrices of the terms
of the matrices of partial derivatives $H_{ij}^{(s)}$ and $C_{jk}^{i(s)}$. From Table 5.7
it directly follows that $^{S}h_{i}^{G(s)} = 0$ for each $i,s \in N$. Translating the
obtained polynomial matrices into polynomial forms, it is easy to obtain the analytical form of the linearized model of cylindrical robot
under consideration. To this end, we will first form the matrix

$$D = \frac{\partial H}{\partial q} \ddot{q} + \dot{q}^{T} \frac{\partial C}{\partial q} \dot{q} + \frac{\partial h_{G}}{\partial q} \, ,$$

and then the complete linearized model (2.5.1). Substituting the partial derivatives from Table 5.8 into the above expression for matrix
D, we obtain

$$D = \begin{bmatrix} 0. & 0. & (0.083 + 8.30 \, x_9)\ddot{q}_1 + 8.30 \, \dot{q}_1\dot{q}_3 \\ 0. & 0. & 0. \\ 0. & 0. & -4.15 \, \dot{q}_1^2 \end{bmatrix}.$$

Substituting into linearized model (2.5.1), it follows directly

$$\delta P_1 = \left(0.40281 + 0.08300 \, x_9 + 4.15000 \, x_9^2\right)\delta\ddot{q}_1 +$$

$$+ 2\left(0.04150 + 4.15000 \, x_9\right)\left(\dot{q}_3\delta\dot{q}_1 + \dot{q}_1\delta\dot{q}_3\right) +$$

$$+ \left(\left(0.083 + 8.300 \, x_9\right)\ddot{q}_1 + 8.300 \, \dot{q}_1\dot{q}_3\right)\delta q_3,$$

$$\delta P_2 = 11.1500 \; \delta \ddot{q}_2,$$

$$\delta P_3 = 4.15000 \; \delta \ddot{q}_3 - 2\left(0.04150 + 4.15000 \; x_9\right)\dot{q}_1 \; \delta \dot{q}_1 - 4.15000 \; \dot{q}_1^2 \delta q_3,$$

where $x_9 = q_3$.

<table>
<tr><td colspan="3" align="center">PARTIAL DERIVATIVE OF INERTIAL MATRIX</td></tr>
<tr><td>i,j</td><td>s</td><td>${}^{S}H_{ij}(s)$</td></tr>
<tr><td>1,1</td><td>3</td><td>$\left(\begin{bmatrix} 0.08300 \\ 8.30000 \end{bmatrix}, \begin{bmatrix} 000 \; 000 \; 000 \\ 000 \; 000 \; 001 \end{bmatrix}\right)$</td></tr>
<tr><td colspan="3" align="center">PARTIAL DERIVATIVE OF C - MATRIX</td></tr>
<tr><td>i</td><td>k,ℓ</td><td>s</td><td>${}^{S}C_{k\ell}^{i}(s)$</td></tr>
<tr><td>1</td><td>1,3</td><td>3</td><td>([4.15000], [000 000 000])</td></tr>
<tr><td>1</td><td>3,1</td><td>3</td><td>([4.15000], [000 000 000])</td></tr>
<tr><td>3</td><td>1,1</td><td>3</td><td>([-4.15000], [000 000 000])</td></tr>
</table>

Table 5.8. Polynomial matrices corresponding to partial derivatives
of terms of dynamic model matrices for cylindrical robot

Construction of sensitivity model of cylindrical robot

The algorithm for forming the sensitivity model was presented in Para.
2.6 and summed in Figs. 2.11 and 2.12. The corresponding model in
numeric-symbolic domain was described in Para. 3.6.

In this example we will assume that the variable dynamic parameters of
the third link are

$$\Delta \theta_3 = \left\{ \Delta m_3, \; \Delta J_{31}, \; \Delta J_{32}, \; \Delta J_{33} \right\},$$

and determine the corresponding sensitivity functions. To this end, we
should form the sensitivity matrices

$$H_s^\mu(q), \quad C_s^\mu(q) \quad \text{and} \quad h_s^{G\mu}(q)$$

where: $s=3$ is serial number of the link with variable parameters, $\mu=0$ corresponds to Δm_3, and $\mu=1,2,3$ corresponds to the variation in parameters $\Delta J_{s\mu}$. The elements of the sensitivity matrices of inertial matrix, calculated according to (3.6.7), are shown in Table 5.9.

i,j	s	μ	$^S H_{sij}^\mu$
1,1	3	0	$\left(\begin{bmatrix} 0.020 \\ 1.000 \end{bmatrix}, \begin{bmatrix} 000\ 000\ 001 \\ 000\ 000\ 002 \end{bmatrix} \right)$
2,2	3	0	$([1.000], [000\ 000\ 000])$
3,3	3	0	$([1.000], [000\ 000\ 000])$
1,1	3	3	$([1.000], [000\ 000\ 000])$

Table 5.9. Polynomial matrices of nonzero elements of sensitivity matrices of the inertial matrix of cylindrical robot

Applying (3.6.8) and (3.6.9), we obtain that $C_{sk\ell}^{i\mu} = 0$, $\{i,k,\ell\} \in N^3$, $s=3$, $\mu \in \{0,1,2,3\}$, except for the elements

$$C_{313}^{1,0} = C_{331}^{1,0} = -C_{311}^{3,0}.$$

The polynomial matrices of these elements are

$$^S C_{313}^{1,0} = \left(\begin{bmatrix} 0.010 \\ 1.000 \end{bmatrix}, \begin{bmatrix} 000\ 000\ 000 \\ 000\ 000\ 001 \end{bmatrix} \right).$$

From (3.6.10) we obtain that

$$^S h_{si}^{G\mu} = 0, \quad s = 3, \quad i \in N,$$

except for $i = 2$ and $\mu = 0$. In this case we obtain

$$^S h_{32}^{G0} = ([9.81], [000\ 000\ 000]).$$

All elements of sensitivity model (2.6.20) - (2.6.22) have thus been formed. Parameters J_{31} and J_{32} have been shown not to affect system dynamics.

Optimization of analytical model computation for cylindrical robot and program code generation

Let us consider the dynamic model of the mechanism of cylindrical robot. The polynomial matrices describing this model are given in Tables 5.5 and 5.6. Although the structure of these matrices is very simple, to illustrate the algorithm developed in Chap. 4, in the text to follow we will present the results of optimizing the computation of the elements of the matrices of dynamic model. First, let us consider polynomial matrices $S_{H_{ij}}$ given in Table 5.5. For each of these matrices we generate the sequences of optimal calculation which are, in matrix form, represented by two matrices

$$E_{H_{ij}} \quad \text{and} \quad V_{H_{ij}}$$

with matrix E containing the nodes and matrix V the branches of the optimal elementary graph (Chapter 4). The results are summed in Table 5.10. On the basis of these results, it is easy to draw the optimal

i,j	$E_{H_{ij}}$	$X_k, \ k>9$	$V_{H_{ij}}$
1,1	$\begin{bmatrix} 9 & 9 & 12 \\ 10 & 11 & 0 \end{bmatrix}$	$X_{10} = 0.40281$ $X_{11} = 0.08300$ $X_{12} = 4.15000$	$\begin{bmatrix} 2 & 1 & 0 \\ 0 & 0 & 0 \end{bmatrix}$
2,2	$\begin{bmatrix} 10 & 0 \end{bmatrix}$	$X_{10} = 11.1500$	$\begin{bmatrix} 0 & 0 \end{bmatrix}$
3,3	$\begin{bmatrix} 10 & 0 \end{bmatrix}$	$X_{10} = 4.15000$	$\begin{bmatrix} 0 & 0 \end{bmatrix}$

Table 5.10. Matrix representatives of the elementary graphs of the inertial matrix of cylindrical robot

graphs. For example, for element H_{11} the following graph is obtained

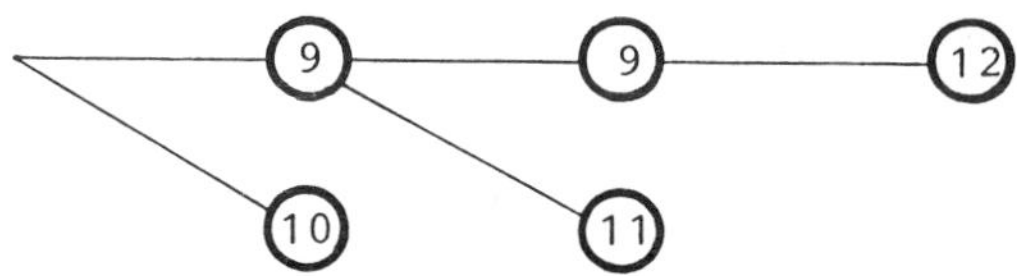

where $X_9 = q_3$.

Let us now consider the generation of the optimal sequence for calculating C - matrices of the system. Since the elements of matrices C^i, $(i=1,2,3)$, may be expressed in terms of elements C_{13}^1, we will give only the matrix representatives of this element:

$$E_{C_{13}^1} = \begin{bmatrix} 9 & 11 \\ 10 & 0 \end{bmatrix}, \quad V_{C_{13}^1} = \begin{bmatrix} 1 & 0 \\ 0 & 0 \end{bmatrix}$$

where $X_{10} = 0.04150$, and $X_{11} = 4.15000$. The optimal sequences of elements h_i^G are trivial, since these elements are 0 or constants.

Program code for real-time calculation of the matrices of dynamic model has been formed according to the theorem from Para. 4.3. The program code has been generated using the expert-program in FORTRAN. The expert program automatically generates the following lines as well:

1) SUBROUTINE <name>

2) COMMON - block,

3) final instructions of RETURN and END type,

where <name> is the desired name of subroutine communicated to the expert-program by a user. Inputs to the subroutine are the joint angles and the joint velocities of robot, which are received through the COMMON-block /UG/. Outputs of the subroutine are the matrices of dynamic model

$$H(q, \theta) \quad \text{and} \quad h(q, \dot{q}, \theta)$$

These data are stored in the COMMON-block /MATRIC/.

Thus, the source program obtained by the expert-program is ready for compilation (by an appropriate compiler). Listing of this program is given in Fig. 5.4.

```
SUBROUTINE UMS2
COMMON/MATRIC/H(6,6),C(6,6,6),G(6)
COMMON/UG/Q(6)
 X7  = Q (1  )
 X8  = Q (2  )
 X9  = Q (3  )
G(3)=0.
G(2)=        0.10938E+03
G(1)=0.
C(3,2,2)=0.
C(3,2,1)=0.
Q1=X9*(-0.41500E+01)
C(3,1,1)=Q1  -0.41500E-01D
C(2,3,3)=0.
C(2,1,1)=0.
C(1,3,3)=0.
C(1,3,2)=0.
C(1,2,2)=0.
H(3,3)=       0.41500E+01
H(3,2)=0.
H(3,1)=0.
H(2,2)=       0.11150E+02
H(2,1)=0.
P1=X9* 0.41500E+01
Q1=X9*(P1+ 0.83000E-01)
H(1,1)=Q1+ 0.40281E+00
```

Fig. 5.4. Program for forming the matrices of the dynamic
          model of cylindrical robot, obtained as output
          of the expert-program

Analysis of the program given in Fig. 5.4 shows that, in spite of the
procedure for optimizing the number of calculations by which the program
has been obtained, there exists some redundancy of operations. Namely,
multiplication in the 12th program line is identical with multiplication in the 24th line. This redundancy results from the fact that the
optimal sequences of calculation are independently generated for different terms of matrices H and C. However, if the algorithm took into ac
account that some expressions in different terms are equal, this would extremely increase its complexity and the generation time of output code.
Of course, the properties of symmetry and antisymmetry, derived in Para.
2.4, are automatically taken into account, and this reduces the redundancy of the number of operations.

Finally, from Fig. 5.4., by simple enumeration we obtain the number of
floating point multiplications n_M and additions n_A

$$n_M = 3, \quad n_A = 3.$$

The execution time of this program on PDP 11/70 minicomputer is approximately 100μs. Thus, the computation of the matrices of the dynamic model

of cylindrical robot may also be achieved using microcomputers of very modest capabilities.

5.3. An Arthropoid Robot

In this example we will consider an arthropoid-structure manipulator. An example of such a robot is given in Fig. 5.5.

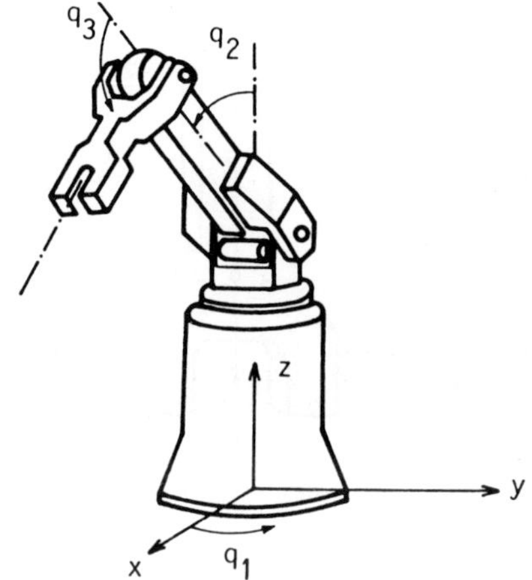

Fig. 5.5. Arthropoid robot

As in the preceding example, the structure of the mechanism of such a robot will be represented by a corresponding model shown in Fig. 5.6.

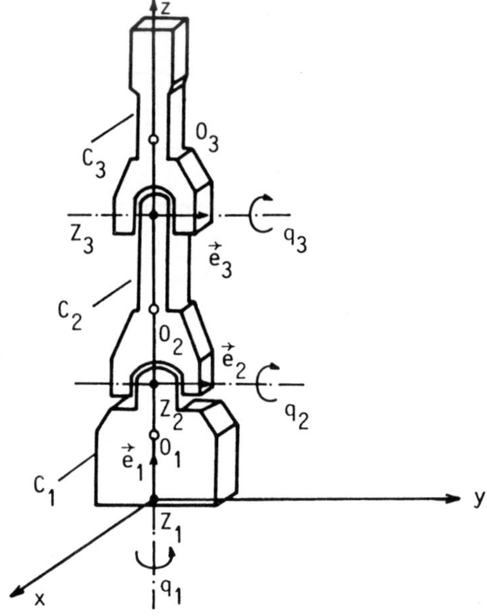

Fig. 5.6. Model of the mechanism of arthropoid robot

The links of mechanism are denoted by C_1, C_2 and C_3, and the joint connections by Z_1, Z_2, and Z_3. Concerning the three rotational degrees of freedom, this mechanism is of RRR structure. The links of mechanism with all geometrical parameters are separately presented in Fig. 5.7.

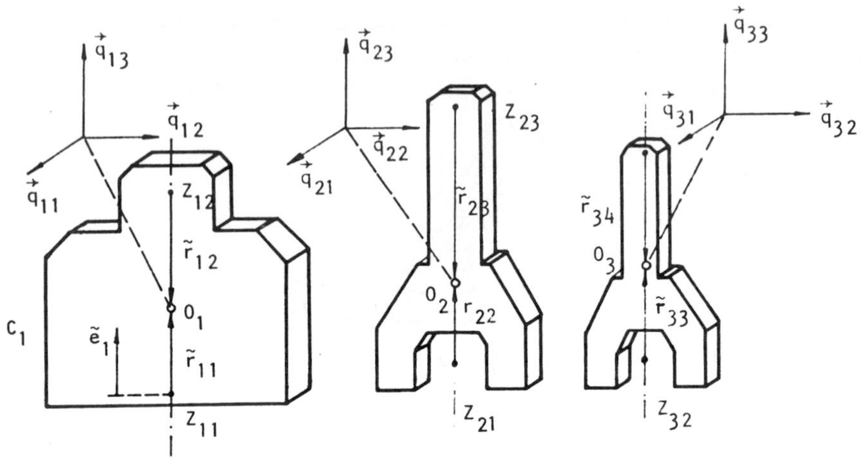

Fig. 5.7. Links of the mechanism of arthropoid robot

Data which determine the kinematic parameters of mechanism links $\tilde{R}_i = \left\{ \tilde{r}_{ii}, \tilde{r}_{i,i+1} \right\}$ and $\tilde{E}_i = \left\{ \tilde{e}_i \right\}$ for $i \in N = \{1,2,3\}$, are given in Table 5.11.

i	$\tilde{r}_{ii}$ [m]	$\tilde{r}_{i,i+1}$ [m]	$\tilde{e}_i$	ξ_i
1	(0., 0., 0.25)	(0., 0., -0.25)	(0., 0., 1.)	0
2	(0., 0., 0.55)	(0., 0., -0.45)	(0., 1., 0.)	0
3	(0., 0., 0.20)	(0., 0., -0.618)	(0., 1., 0.)	0

Table 5.11. Kinematic parameters of arthropoid robot

The dynamic parameters of the mechanism are shown in Table 5.12. The parameters which are not included in this table are not important for constructing the dynamic model of arthropoid robot.

i	m_i [kg]	J_{i1} [kg m^2]	J_{i2} [kg m^2]	J_{i3} [kg m^2]
1	–	–	–	4.57
2	152.	–	21.74	–
3	67.	–	1.22	–

Table 5.12. Dynamic parameters of arthropoid robot

Derivation of nonlinear analytical model of
arthropoid robot

The first stage of the algorithm for forming the nonlinear analytical
model (see Fig. 2.7) includes determination of the "home" position of
mechanism. The output of this stage is

$$Q_i^O = \begin{bmatrix} 1 & 0 & 0 \\ 0 & 1 & 0 \\ 0 & 0 & 1 \end{bmatrix},$$

$$\vec{r}_{ii}^{\,O} = \tilde{r}_{ii}, \quad \vec{r}_{i,i+1}^{\,O} = \tilde{r}_{i,i+1}, \qquad (i=1,2,3),$$

and

$$\vec{e}_{i+1}^{\,O} = \tilde{e}_{i+1}, \qquad (i=1,2).$$

In the next stage of the algorithm, the position of the links of the
mechanism as a function of joint coordinates q_1, q_2 and q_3 is determined. The output of this stage represents the polynomial matrices

$$S_{q_{ij}}, \quad S_{r_{ii}}, \quad S_{r_{i,i+1}}, \quad S_{e_i}, \qquad i \in N, \quad j=1,2,3.$$

Computer generation of these matrices yields data presented in Tables
5.13 and 5.14. In these tables the matrices of exponents have only 6
columns, since the 7th, 8th and 9th columns are, obviously, to be zero
vectors. Evidently, this assumption holds only when the system has no
sliding joints.

In the next stage of the algorithm, the polynomial matrices of the
terms of the robot mechanism inertial matrix are formed according to

i	j	$^{S}q_{ij}$
1	1	$\left(\begin{bmatrix} 1. & 0. & 0. \\ 0. & 1. & 0. \end{bmatrix} , \begin{bmatrix} 1 & 0 & 0 & 0 & 0 & 0 \\ 0 & 0 & 0 & 1 & 0 & 0 \end{bmatrix} \right)$
	2	$\left(\begin{bmatrix} 0. & 1. & 0. \\ -1. & 0. & 0. \end{bmatrix} , \begin{bmatrix} 1 & 0 & 0 & 0 & 0 & 0 \\ 0 & 0 & 0 & 1 & 0 & 0 \end{bmatrix} \right)$
	3	$\left(\begin{bmatrix} 0. & 0. & 1. \end{bmatrix} , \begin{bmatrix} 0 & 0 & 0 & 0 & 0 & 0 \end{bmatrix} \right)$
2	1	$\left(\begin{bmatrix} 1. & 0. & 0. \\ 0. & 1. & 0. \\ 0. & 0. & -1. \end{bmatrix} , \begin{bmatrix} 1 & 1 & 0 & 0 & 0 & 0 \\ 0 & 1 & 0 & 1 & 0 & 0 \\ 0 & 0 & 0 & 0 & 1 & 0 \end{bmatrix} \right)$
	2	$\left(\begin{bmatrix} 0. & 1. & 0. \\ -1. & 0. & 0. \end{bmatrix} , \begin{bmatrix} 1 & 0 & 0 & 0 & 0 & 0 \\ 0 & 0 & 0 & 1 & 0 & 0 \end{bmatrix} \right)$
	3	$\left(\begin{bmatrix} 0. & 0. & 1. \\ 1. & 0. & 0. \\ 0. & 1. & 0. \end{bmatrix} , \begin{bmatrix} 0 & 1 & 0 & 0 & 0 & 0 \\ 1 & 0 & 0 & 0 & 1 & 0 \\ 0 & 0 & 0 & 1 & 1 & 0 \end{bmatrix} \right)$
3	1	$\left(\begin{bmatrix} 1. & 0. & 0. \\ 0. & 0. & -1. \\ 0. & 1. & 0. \\ 0. & 0. & -1. \\ -1. & 0. & 0. \\ 0. & 1. & 0. \end{bmatrix} , \begin{bmatrix} 1 & 1 & 1 & 0 & 0 & 0 \\ 0 & 0 & 1 & 0 & 1 & 0 \\ 0 & 1 & 1 & 1 & 0 & 0 \\ 0 & 1 & 0 & 0 & 0 & 1 \\ 1 & 0 & 0 & 0 & 1 & 1 \\ 0 & 0 & 0 & 1 & 1 & 1 \end{bmatrix} \right)$
	2	$\left(\begin{bmatrix} 0. & 1. & 0. \\ -1. & 0. & 0. \end{bmatrix} , \begin{bmatrix} 1 & 0 & 0 & 0 & 0 & 0 \\ 0 & 0 & 0 & 1 & 0 & 0 \end{bmatrix} \right)$

Table 5.13. Polynomial matrices corresponding to the unit vectors of local coordinate frames for arthropoid robot

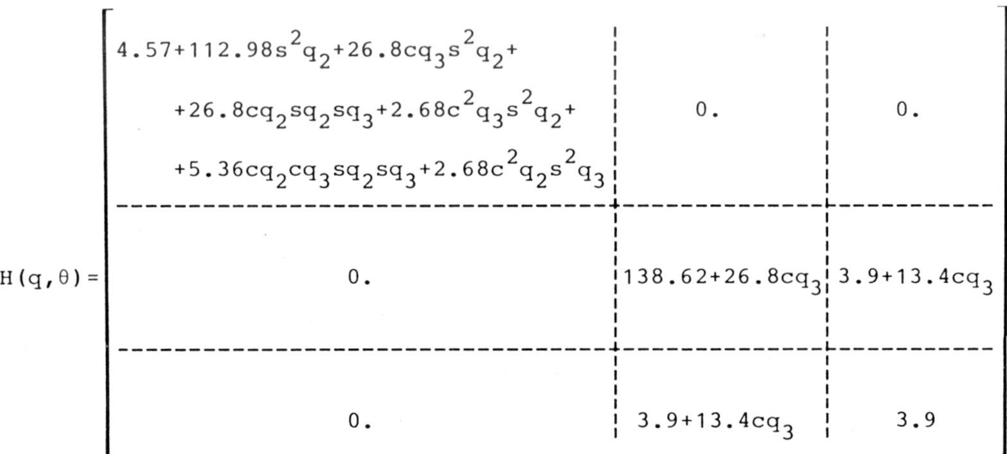

Table 5.13. (Contd.)

equation (3.5.10). The results obtained are presented in Table 5.15.
Since the property of symmetry of the inertial matrix holds (see Para.
2.4), Table 5.15 contains elements H_{ij}, $i < j$. As may be seen from
the table, elements $H_{12} = H_{21}$ and $H_{13} = H_{31}$ are zero elements. This
means that the first degree of freedom (rotation about vertical axis)
is inertially "decoupled" with respect to the second and third degrees
of freedom (rotations about horizontal axes). Since $H_{33} = 3.9 = \text{const.}$,
the third degree of freedom is "loaded" by a constant inertial load.

For illustration, let us write matrix $H(q, 0) \in R^{3 \times 3}$ which corresponds
to data given in Table 5.15:

$$H(q,\theta) = \begin{bmatrix} \begin{matrix} 4.57+112.98s^2q_2+26.8cq_3s^2q_2+ \\ +26.8cq_2sq_2sq_3+2.68c^2q_3s^2q_2+ \\ +5.36cq_2cq_3sq_2sq_3+2.68c^2q_2s^2q_3 \end{matrix} & 0. & 0. \\ 0. & 138.62+26.8cq_3 & 3.9+13.4cq_3 \\ 0. & 3.9+13.4cq_3 & 3.9 \end{bmatrix}$$

where cq_i and sq_i denote $\cos q_i$ and $\sin q_i$ respectively.

Numeric-symbolic forms of the elements of C – matrices of the model are
generated in the next stage of the algorithm according to equation

i	S_{e_i}		
1	$([0.\quad 0.\quad 1.]\ ,\ [0\ 0\ 0\quad 0\ 0\ 0])$		
2	$\left(\begin{bmatrix} 0. & 1. & 0. \\ -1. & 0. & 0. \end{bmatrix},\ \begin{bmatrix} 1\ 0\ 0 & 0\ 0\ 0 \\ 0\ 0\ 0 & 1\ 0\ 0 \end{bmatrix} \right)$		
3	$\left(\begin{bmatrix} 0. & 1. & 0. \\ -1. & 0. & 0. \end{bmatrix},\ \begin{bmatrix} 1\ 0\ 0 & 0\ 0\ 0 \\ 0\ 0\ 0 & 1\ 0\ 0 \end{bmatrix} \right)$		

i	r_{ii}		
1	$([0.\quad 0.\quad 0.25],\ [0\ 0\ 0\quad 0\ 0\ 0])$		
2	$\left(\begin{bmatrix} 0.55 & 0. & 0. \\ 0. & 0.55 & 0. \\ 0. & 0. & 0.55 \end{bmatrix},\ \begin{bmatrix} 1\ 0\ 0 & 0\ 1\ 0 \\ 0\ 0\ 0 & 1\ 1\ 0 \\ 0\ 1\ 0 & 0\ 0\ 0 \end{bmatrix} \right)$		
3	$\left(\begin{bmatrix} 0. & 0. & 0.2 \\ 0.2 & 0. & 0. \\ 0. & 0.2 & 0. \\ 0.2 & 0. & 0. \\ 0. & 0.2 & 0. \\ 0. & 0. & -0.2 \end{bmatrix},\ \begin{bmatrix} 0\ 1\ 1 & 0\ 0\ 0 \\ 1\ 0\ 1 & 0\ 1\ 0 \\ 0\ 0\ 1 & 1\ 1\ 0 \\ 1\ 1\ 0 & 0\ 0\ 1 \\ 0\ 1\ 0 & 1\ 0\ 1 \\ 0\ 0\ 0 & 0\ 1\ 1 \end{bmatrix} \right)$		

i	$S_{r_{i,i+1}}$		
1	$([0.\quad 0.\quad -0.25],\ [0\ 0\ 0\quad 0\ 0\ 0])$		
2	$\left(\begin{bmatrix} 0. & 0. & -0.45 \\ -0.45 & 0. & 0. \\ 0. & -0.45 & 0. \end{bmatrix},\ \begin{bmatrix} 0\ 1\ 0 & 0\ 0\ 0 \\ 1\ 0\ 0 & 0\ 1\ 0 \\ 0\ 0\ 0 & 1\ 1\ 0 \end{bmatrix} \right)$		

Table 5.14. Polynomial matrices S_{e_i}, $S_{r_{ii}}$ and $S_{r_{i,i+1}}$ of arthropoid robot

i,j	$S_{H_{ij}}$
1,1	$\left(\begin{bmatrix} 4.57000 \\ 112.98000 \\ 26.80000 \\ 26.80000 \\ 2.68000 \\ 5.36000 \\ 2.68000 \end{bmatrix} , \begin{bmatrix} 0 & 0 & 0 & 0 & 0 & 0 \\ 0 & 0 & 0 & 0 & 2 & 0 \\ 0 & 0 & 1 & 0 & 2 & 0 \\ 0 & 1 & 0 & 0 & 1 & 1 \\ 0 & 0 & 2 & 0 & 2 & 0 \\ 0 & 1 & 1 & 0 & 1 & 1 \\ 0 & 2 & 0 & 0 & 0 & 2 \end{bmatrix} \right)$
1,2	$([\ 0.00000]\ ,\ [0\ 0\ 0\ \ 0\ 0\ 0])$
1,3	$([\ 0.00000]\ ,\ [0\ 0\ 0\ \ 0\ 0\ 0])$
2,2	$\left(\begin{bmatrix} 138.62001 \\ 26.80000 \end{bmatrix} , \begin{bmatrix} 0 & 0 & 0 & 0 & 0 & 0 \\ 0 & 0 & 1 & 0 & 0 & 0 \end{bmatrix} \right)$
2,3	$\left(\begin{bmatrix} 3.90000 \\ 13.40000 \end{bmatrix} , \begin{bmatrix} 0 & 0 & 0 & 0 & 0 & 0 \\ 0 & 0 & 1 & 0 & 0 & 0 \end{bmatrix} \right)$
3,3	$([\ 3.90000\ ]\ ,\ [0\ 0\ 0\ \ 0\ 0\ 0])$

Table 5.15. Polynomial matrices of the elements of the inertial matrix of arthropoid robot

(3.5.11). Nonzero elements of C - matrices are given in Table 5.16. As may be seen, all elements of C - matrices may be expressed by four elements c_{21}^1, c_{31}^1, c_{32}^2 and c_{33}^2. All the remaining elements are obtained using the property of symmetry and antisymmetry. So, matrices C^i, $i \in N$ become

$$C^1 = \begin{bmatrix} 0 & c_{21}^1 & c_{31}^1 \\ c_{21}^1 & 0 & 0 \\ c_{31}^1 & 0 & 0 \end{bmatrix} ,$$

184

$$C^2 = \begin{bmatrix} -C_{21}^1 & 0 & 0 \\ 0 & 0 & C_{32}^2 \\ 0 & C_{32}^2 & C_{32}^2 \end{bmatrix}$$

$$C^3 = \begin{bmatrix} -C_{31}^1 & 0 & 0 \\ 0 & -C_{32}^2 & 0 \\ 0 & 0 & 0 \end{bmatrix} .$$

i	k, ℓ	$^S C_{k\ell}^i$
1	2,1	$\left(\begin{bmatrix} 112.980 \\ 26.800 \\ 13.400 \\ 2.680 \\ 2.680 \\ -13.400 \\ -2.680 \\ -2.680 \end{bmatrix} , \begin{bmatrix} 0 & 1 & 0 & 0 & 1 & 0 \\ 0 & 1 & 1 & 0 & 1 & 0 \\ 0 & 2 & 0 & 0 & 0 & 1 \\ 0 & 1 & 2 & 0 & 1 & 0 \\ 0 & 2 & 1 & 0 & 0 & 1 \\ 0 & 0 & 0 & 0 & 2 & 1 \\ 0 & 0 & 1 & 0 & 2 & 1 \\ 0 & 1 & 0 & 0 & 1 & 2 \end{bmatrix} \right)$
1	3,1	$\left(\begin{bmatrix} 13.400 \\ 2.680 \\ 2.680 \\ -13.400 \\ -2.680 \\ -2.680 \end{bmatrix} , \begin{bmatrix} 0 & 1 & 1 & 0 & 1 & 0 \\ 0 & 1 & 2 & 0 & 1 & 0 \\ 0 & 2 & 1 & 0 & 0 & 1 \\ 0 & 0 & 0 & 0 & 2 & 1 \\ 0 & 0 & 1 & 0 & 2 & 1 \\ 0 & 1 & 0 & 0 & 1 & 2 \end{bmatrix} \right)$
2	3,2	$([-13.400], [0\ 0\ 0\ \ 0\ 0\ 1])$
2	3,3	$([-13.400], [0\ 0\ 0\ \ 0\ 0\ 1])$

Table 5.16. Polynomial matrices of nonzero elements of
C - matrices of arthropoid robot

The polynomial matrices of gravity vector $S_{h_i^G}$, $i \in N$ are calculated in the last stage of the algorithm according to equation (3.5.12). Thus, we obtain

$$S_{h_1^G} = 0$$

$$S_{h_2^G} = \left(\begin{bmatrix} -1477.3861 \\ -131.4540 \\ -131.4540 \end{bmatrix} , \begin{bmatrix} 0 & 0 & 0 & 0 & 1 & 0 \\ 0 & 0 & 1 & 0 & 1 & 0 \\ 0 & 1 & 0 & 0 & 0 & 1 \end{bmatrix} \right)$$

$$S_{h_3^G} = \left(\begin{bmatrix} -131.4540 \\ -131.4540 \end{bmatrix} , \begin{bmatrix} 0 & 0 & 1 & 0 & 1 & 0 \\ 0 & 1 & 0 & 0 & 0 & 1 \end{bmatrix} \right).$$

Thus, the polynomial matrices corresponding to all elements of dynamic model matrices for arthropoid robot have been determined.

Construction of linearized model of arthropoid robot

To obtain the linearized model of robot in a numeric-symbolic form, it is necessary to determine the polynomial matrices

$$S_{H_{ij}}(s), \quad S_{C_{jk}^i}(s) \quad \text{and} \quad S_{h_i^G}(s)$$

for $i,j,k,s \in N$, $(s) \equiv \partial/\partial q_s$. To this end, it is possible to apply the theorem of the partial derivative of polynomial matrix, presented in Para. 3.6. Implementing the algorithm described in this theorem, a program is obtained for computer-aided generation of the linearized model. The polynomial matrices corresponding to the partial derivatives of the terms of the inertial matrix of arthropoid robot are given in Table 5.17. The table contains only the partial derivatives which are different from zero. As may be seen, the matrices of the partial derivatives of the inertial matrix of arthropoid robot have the following structure

$$\frac{\partial H(q, \theta)}{\partial q_2} = \begin{bmatrix} H_{11}^{(2)} & 0 & 0 \\ 0 & 0 & 0 \\ 0 & 0 & 0 \end{bmatrix} ,$$

$$\frac{\partial H(q,\ \theta)}{\partial q_3} = \begin{bmatrix} H_{11}^{(3)} & 0 & 0 \\ 0 & H_{22}^{(3)} & H_{23}^{(3)} \\ 0 & H_{32}^{(3)} & 0 \end{bmatrix}$$

i,j	s	${}^{S}H_{ij}^{(s)}$
1,1	2	$\left(\begin{bmatrix} 225.960 \\ 53.600 \\ 26.800 \\ -53.600 \\ 5.360 \\ 5.360 \\ -10.720 \\ -2.680 \end{bmatrix},\ \begin{bmatrix} 0 & 1 & 0 & 0 & 1 & 0 \\ 0 & 1 & 1 & 0 & 1 & 0 \\ 0 & 0 & 0 & 0 & 0 & 1 \\ 0 & 0 & 0 & 0 & 2 & 1 \\ 0 & 1 & 2 & 0 & 1 & 0 \\ 0 & 0 & 1 & 0 & 0 & 1 \\ 0 & 0 & 1 & 0 & 2 & 1 \\ 0 & 1 & 0 & 0 & 1 & 2 \end{bmatrix}\right)$
1,1	3	$\left(\begin{bmatrix} -26.800 \\ 26.800 \\ -5.360 \\ 5.360 \\ -10.720 \\ 5.360 \end{bmatrix},\ \begin{bmatrix} 0 & 0 & 0 & 0 & 2 & 1 \\ 0 & 1 & 1 & 0 & 1 & 0 \\ 0 & 0 & 1 & 0 & 2 & 1 \\ 0 & 1 & 0 & 0 & 1 & 0 \\ 0 & 1 & 0 & 0 & 1 & 2 \\ 0 & 2 & 1 & 0 & 0 & 1 \end{bmatrix}\right)$
2,2	3	$([-26.800],\ [0\ 0\ 0\quad 0\ 0\ 1])$
2,3	3	$([-13.400],\ [0\ 0\ 0\quad 0\ 0\ 1])$

Table 5.17. Polynomial matrices of the partial derivatives of the terms of the inertial matrix for arthropoid robot

The partial derivatives of the terms of C - matrices of dynamic model are given in Table 5.18. It may be seen that the matrices of the partial derivatives of C - matrices of arthropoid robot contain a large number of zero elements, as was the case with the inertial matrix.

Polynomial matrices of the partial derivatives of the elements of

i	k,ℓ	s	$^{S}C^{i(s)}_{k\ell}$		
1	2,1	2	$\left(\begin{bmatrix} 115.660 \\ -231.320 \\ 26.800 \\ -53.600 \\ -53.600 \\ -5.360 \\ 10.720 \end{bmatrix}\right.$	,	$\left.\begin{bmatrix} 0\ 0\ 0 & 0\ 0\ 0 \\ 0\ 0\ 0 & 0\ 2\ 0 \\ 0\ 0\ 1 & 0\ 0\ 0 \\ 0\ 0\ 1 & 0\ 2\ 0 \\ 0\ 1\ 0 & 0\ 1\ 1 \\ 0\ 0\ 0 & 0\ 0\ 2 \\ 0\ 0\ 0 & 0\ 2\ 2 \end{bmatrix}\right)$
1	2,1	3	$\left(\begin{bmatrix} -26.800 \\ 13.400 \\ -10.720 \\ 2.680 \\ -5.360 \\ -13.400 \\ -5.360 \\ 10.720 \end{bmatrix}\right.$	,	$\left.\begin{bmatrix} 0\ 1\ 0 & 0\ 1\ 1 \\ 0\ 2\ 1 & 0\ 0\ 0 \\ 0\ 1\ 1 & 0\ 1\ 1 \\ 0\ 0\ 0 & 0\ 0\ 0 \\ 0\ 0\ 0 & 0\ 0\ 2 \\ 0\ 0\ 1 & 0\ 2\ 0 \\ 0\ 0\ 0 & 0\ 2\ 0 \\ 0\ 0\ 0 & 0\ 2\ 2 \end{bmatrix}\right)$
1	3,1	2	$\left(\begin{bmatrix} 13.400 \\ -26.800 \\ 2.680 \\ -5.360 \\ -5.360 \\ 10.720 \\ -26.800 \end{bmatrix}\right.$	,	$\left.\begin{bmatrix} 0\ 0\ 1 & 0\ 0\ 0 \\ 0\ 0\ 1 & 0\ 2\ 0 \\ 0\ 0\ 0 & 0\ 0\ 0 \\ 0\ 0\ 0 & 0\ 0\ 2 \\ 0\ 0\ 0 & 0\ 2\ 0 \\ 0\ 0\ 0 & 0\ 2\ 2 \\ 0\ 1\ 0 & 0\ 1\ 1 \end{bmatrix}\right)$

Table 5.18. Polynomial matrices of the partial derivatives of the terms of C - matrices for arthropoid robot

1	3,1	3	$\left(\begin{bmatrix} -13.400 \\ -10.720 \\ 2.720 \\ -5.360 \\ -13.400 \\ -5.360 \end{bmatrix}, \begin{bmatrix} 0 & 1 & 0 & 0 & 1 & 1 \\ 0 & 1 & 1 & 0 & 1 & 1 \\ 0 & 0 & 0 & 0 & 0 & 0 \\ 0 & 2 & 0 & 0 & 0 & 2 \\ 0 & 0 & 1 & 0 & 2 & 0 \\ 0 & 0 & 2 & 0 & 2 & 0 \end{bmatrix} \right)$
2	3,2	3	$([-13.400], \; [0 \; 0 \; 1 \; 0 \; 0 \; 0])$
2	3,3	3	$([-13.400], \; [0 \; 0 \; 1 \; 0 \; 0 \; 0])$

Table 5.18. (Contd.)

gravity vector are formed in the last stage of linearized model construction. Thus, one obtains

$$S \frac{\partial h_2^G}{\partial q_2} = \left(\begin{bmatrix} -1477.3861 \\ -131.4540 \\ 131.4540 \end{bmatrix}, \begin{bmatrix} 0 & 1 & 0 & 0 & 0 & 0 \\ 0 & 1 & 1 & 0 & 0 & 0 \\ 0 & 0 & 0 & 0 & 1 & 1 \end{bmatrix} \right)$$

$$S \frac{\partial h_2^G}{\partial q_3} = \left(\begin{bmatrix} 131.4540 \\ -131.4540 \end{bmatrix}, \begin{bmatrix} 0 & 0 & 0 & 0 & 1 & 1 \\ 0 & 1 & 1 & 0 & 0 & 0 \end{bmatrix} \right)$$

$$S \frac{\partial h_3^G}{\partial q_2} = \left(\begin{bmatrix} -131.4540 \\ 131.4540 \end{bmatrix}, \begin{bmatrix} 0 & 1 & 1 & 0 & 0 & 0 \\ 0 & 0 & 0 & 0 & 1 & 1 \end{bmatrix} \right)$$

$$S \frac{\partial h_3^G}{\partial q_3} = \left(\begin{bmatrix} 131.4540 \\ -131.4540 \end{bmatrix}, \begin{bmatrix} 0 & 0 & 0 & 0 & 1 & 1 \\ 0 & 1 & 1 & 0 & 0 & 0 \end{bmatrix} \right)$$

Derivation of the matrices of linearized model is completed in this way.

Construction of the sensitivity model of arthropoid robot

In this section we will illustrate the method for forming the sensitivity model (see Para. 3.6) using the example of arthropoidal robot. In
the case of cylindrical robot we have assumed that the parameters of

the third link are variable. Here we will assume that the parameters of the second link are variable. This assumption is possible because the algorithm for sensitivity model construction does not depend upon the serial number of the link with variable parameters.

Let us consider the set of parameters

$$\theta_2 = \left\{ m_2, \ J_{21}, \ J_{22}, \ J_{23} \right\}$$

which is assumed to vary. We will form the sensitivity matrices

$$H_s^{\mu}(q), \quad C_s^{\mu}(q) \quad \text{and} \quad h_s^{\mu}(q)$$

where: $s = 2$ is serial number of the link with variable parameters, $\mu = 0$ corresponds to parameter m_2, $\mu = 1,2,3$ to parameters $\Delta J_{s\mu}$. According to (3.6.7) we obtain the data given in Table 5.19. It may be seen from the table that in this case the sensitivity model is considerably simpler than the complete nonlinear model. It may also be seen that since parameters J_{21} and J_{23} are not included into the model matrices, the sensitivity to variations of these parameters equals zero.

INERTIAL MATRIX			
i,j	s	μ	$S_H{}^{i\mu}_{sij}$
1,1	2	0	([0.3025], [0 0 0 0 2 0])
2,2	2	0	([0.3025], [0 0 0 0 0 0])
2,2	2	2	([1.0000], [0 0 0 0 0 0])
C - MATRIX			
i,k,ℓ	s	μ	$S_H{}^{i\mu}_{sk\ell}$
1,2,1	2	0	([0.3025], [0 1 0 0 1 0])
GRAVITY VECTOR			
i	s	μ	$S_h{}^{G\mu}_{sk}$
2	2	0	([5.39551], [0 0 0 0 1 0])

Table 5.19. Polynomial matrices of nonzero elements of the
sensivity matrices for arthropoid robot

Construction of approximate models of arthropoid robot

A method for forming approximate robot models with the possibility for prescribing the admissible error was presented in Para. 3.7. According to this procedure, we first perform normalization by transition from the domain of arguments $X = \{x_1, \ldots, x_N\}$ into domain $Z = \{z_1, \ldots, z_N\}$ through linear mapping $z_i = a_i x_i$, where a_i is determined by

$$a_i = 1 \quad \text{for} \quad i=1,\ldots,2n$$

$$a_i = \sup_t |q_i(t)|, \quad i=2n+1,\ldots,3n.$$

Since the case of arthropoid robot involves rotational degrees of freedom, and no translational joints, the matrices of dynamic model are independent of coordinates $x_{2n+1},\ldots,x_{3n}$. This allows us to accept

$$a_i = 1 \quad \text{for} \quad i=1,\ldots,3n,$$

which corresponds to unit mapping $X \rightarrow X$.

Let us now consider the inertial matrix given by Table 5.15. Normalization of coefficients should be performed. Normalization with respect to the largest element in the row will be accepted (see (3.7.20)). Thus, one obtains

$$S_{H_{11}} = \left(112.98 \begin{bmatrix} 0.0404 \\ 1.0000 \\ 0.2372 \\ 0.2372 \\ 0.0237 \\ 0.0474 \\ 0.0237 \end{bmatrix}, \begin{bmatrix} 0 & 0 & 0 & 0 & 0 & 0 \\ 0 & 0 & 0 & 0 & 2 & 0 \\ 0 & 0 & 1 & 0 & 2 & 0 \\ 0 & 1 & 0 & 0 & 1 & 1 \\ 0 & 0 & 2 & 0 & 2 & 0 \\ 0 & 1 & 1 & 0 & 1 & 1 \\ 0 & 2 & 0 & 0 & 0 & 2 \end{bmatrix} \right)$$

$$S_{H_{22}} = \left(138.62 \begin{bmatrix} 1.0000 \\ 0.1933 \end{bmatrix}, \begin{bmatrix} 0 & 0 & 0 & 0 & 0 & 0 \\ 0 & 0 & 1 & 0 & 0 & 0 \end{bmatrix} \right)$$

$$S_{H_{23}} = \left(138.62 \begin{bmatrix} 0.0281 \\ 0.0967 \end{bmatrix}, \begin{bmatrix} 0 & 0 & 0 & 0 & 0 & 0 \\ 0 & 0 & 1 & 0 & 0 & 0 \end{bmatrix} \right)$$

$$S_{H_{32}} = \left(138.62 \begin{bmatrix} 0.2910 \\ 1.0000 \end{bmatrix}, \quad \begin{bmatrix} 0 & 0 & 0 & 0 & 0 & 0 \\ 0 & 0 & 1 & 0 & 0 & 0 \end{bmatrix}\right)$$

$$S_{H_{33}} = (13.40 \quad [0.2910], \quad [0\ 0\ 0\ \ 0\ 0\ 0])$$

Let the given admissible deviation be 5% in the sense of definition (3.7.12). Now we may apply the theorem from Ch. 3.7. Since $|0.0237|$ + $|0.0237| < 0.05$, it is possible to eliminate the 5th and 7th row from the structural matrix for H_{11}. In the same way we conclude that the first row from polynomial matrix $S_{H_{23}}$ can also be eliminated. Thus, we obtain

$$\tilde{S}_{H_{11}} = \left(112.98 \begin{bmatrix} 0.0404 \\ 1.0000 \\ 0.2372 \\ 0.2372 \\ 0.0474 \end{bmatrix}, \quad \begin{bmatrix} 0 & 0 & 0 & 0 & 0 & 0 \\ 0 & 0 & 0 & 0 & 2 & 0 \\ 0 & 0 & 1 & 0 & 2 & 0 \\ 0 & 1 & 0 & 0 & 1 & 1 \\ 0 & 1 & 1 & 0 & 1 & 1 \end{bmatrix}\right),$$

$$\tilde{S}_{H_{23}} = (138.62 \quad [0.0967], \quad [0\ 0\ 1\ \ 0\ 0\ 0]),$$

$$\tilde{S}_{H_{ij}} = S_{H_{ij}}, \quad (i,j) = \begin{cases} (2,2) \\ (3,2), \\ (3,3) \end{cases}$$

where $\tilde{S}$ denotes the approximate value of polynomial matrix. The driving torques can now be expressed by approximate dynamic equations (see (2.2.17) and (2.3.1))

$$P_1 \simeq \tilde{H}_{11}\ddot{q}_1 + h_1$$

$$P_2 = \tilde{H}_{22}\ddot{q}_2 + \tilde{H}_{23}\ddot{q}_3 + h_2$$

$$P_3 = \tilde{H}_{32}\ddot{q}_2 + \tilde{H}_{33}\ddot{q}_2 + h_3$$

where

$$h_1 = 2c_{21}^1\dot{q}_1\dot{q}_2 + 2c_{31}^1\dot{q}_1\dot{q}_3 + h_1^G$$

$$h_2 = c_{21}^1\dot{q}_1^2 + 2c_{32}^2\dot{q}_2\dot{q}_3 + c_{33}^2\dot{q}_3^2 + h_2^G$$

$$h_3 = c^1_{31}\dot{q}^2_1 + c^2_{32}\dot{q}^2_2 + h^G_3$$

with $c^i_{k\ell}$ $(i,k,\ell \in N)$ given in Table 5.16.

The task is to determine the approximate expressions for h_1, h_2 and h_3 with the allowable deviation not exceeding 5% in the sense of (3.7.12). To achieve this, we should first normalize the expressions for h_i, $i \in N$, to maximum internal velocities $\dot{q}_{i(max)}$. For example,

$$|\dot{q}_1|_{max} = 1 \; \frac{rad}{s}$$

$$|\dot{q}_2|_{max} = |\dot{q}_3|_{max} = 2 \; \frac{rad}{s}.$$

Let us introduce substitution $v_i = \dot{q}_i/|\dot{q}_i|_{max}$, $i \in N$, into expressions for h_1, h_2 and h_3. Thus, we obtain

$$h_1 = 4c^1_{21}v_1v_2 + 4c^1_{31}v_1v_3 + h^G_1$$

$$h_2 = -c^1_{21}v^2_1 + 8c^2_{32}v_2v_3 + 4c^2_{33}v^2_3 + h^G_2$$

$$h_3 = -c^1_{31}v^2_1 - 4c^2_{32}v^2_2 + h^G_3$$

where $|v_i| < 1$.

We will now form the polynomial matrices for h_i, $i \in N$, as follows. Let us introduce the following notation for the elements of C - matrix and gravity vector

$$S_{c^i_{k\ell}} = \left(V^i_{k\ell}, \; E^i_{k\ell} \right), \qquad k, \ell, i \in N$$

$$S_{h^G_i} = \left(V^G_i, \; E^G_i \right), \qquad i \in N$$

where the matrices of exponents are defined according to the arranged set of arguments

$$x = \left(\cos q_1, \; \cos q_2, \; \cos q_3, \; \sin q_1, \; \sin q_2, \; \sin q_3 \right)$$

which will be shorter represented by

$$x = \left(\cos q_i \mid \sin q_i \right).$$

We will extend the set of arguments by elements v_1, v_2 and v_3, and we obtain

$$x_e = \left(\cos q_i \mid \sin q_i \mid v_i\right).$$

h_1, h_2 and h_3 can now be represented by the following polynomial matrices

$$h_1 = \left(\begin{bmatrix} 4V_{21}^1 \\ \hline 4V_{31}^1 \\ \hline V_1^G \end{bmatrix}, \begin{bmatrix} E_{21}^1 & \begin{matrix} 1 & 1 & 0 \\ \cdots & \\ 1 & 1 & 0 \end{matrix} \\ \hline E_{31}^1 & \begin{matrix} 1 & 0 & 1 \\ \cdots & \\ 1 & 0 & 1 \end{matrix} \\ \hline E_1^G & \begin{matrix} 0 & 0 & 0 \\ \cdots & \\ 0 & 0 & 0 \end{matrix} \end{bmatrix}\right),$$

$$h_2 = \left(\begin{bmatrix} -V_{21}^1 \\ \hline 8V_{32}^2 \\ \hline 4V_{33}^2 \\ \hline V_2^G \end{bmatrix}, \begin{bmatrix} E_{21}^1 & \begin{matrix} 2 & 0 & 0 \\ \cdots & \\ 2 & 0 & 0 \end{matrix} \\ \hline E_{32}^2 & \begin{matrix} 0 & 1 & 1 \\ \cdots & \\ 0 & 1 & 1 \end{matrix} \\ \hline E_{33}^2 & \begin{matrix} 0 & 0 & 2 \\ \cdots & \\ 0 & 0 & 2 \end{matrix} \\ \hline E_2^G & \begin{matrix} 0 & 0 & 0 \\ \cdots & \\ 0 & 0 & 0 \end{matrix} \end{bmatrix}\right),$$

$$h_3 = \left(\begin{bmatrix} -V_{31}^1 \\ \hline -4V_{32}^2 \\ \hline V_3^G \end{bmatrix}, \begin{bmatrix} E_{31}^1 & \begin{matrix} 2 & 0 & 0 \\ \cdots & \\ 2 & 0 & 0 \end{matrix} \\ \hline E_{32}^2 & \begin{matrix} 0 & 2 & 0 \\ \cdots & \\ 0 & 2 & 0 \end{matrix} \\ \hline E_3^G & \begin{matrix} 0 & 0 & 0 \\ \cdots & \\ 0 & 0 & 0 \end{matrix} \end{bmatrix}\right),$$

i.e., in the form

$$h_i = \left(V_i, E_i\right), \qquad i \in N.$$

Using the data from Table 5.16, we obtain

$$
S_{h_1} = \left(
\begin{bmatrix}
451.920 \\
107.200 \\
53.600 \\
10.720 \\
10.720 \\
-53.600 \\
-10.720 \\
-10.720 \\
\hline
53.600 \\
10.720 \\
10.720 \\
-53.600 \\
-10.720 \\
-10.720
\end{bmatrix}
,\;
\left[
\begin{array}{cccccc|ccc}
0 & 1 & 0 & 0 & 1 & 0 & 1 & 1 & 0 \\
0 & 1 & 1 & 0 & 1 & 0 & 1 & 1 & 0 \\
0 & 2 & 0 & 0 & 0 & 1 & 1 & 1 & 0 \\
0 & 1 & 2 & 0 & 1 & 0 & 1 & 1 & 0 \\
0 & 2 & 1 & 0 & 0 & 1 & 1 & 1 & 0 \\
0 & 0 & 0 & 0 & 2 & 1 & 1 & 1 & 0 \\
0 & 0 & 1 & 0 & 2 & 1 & 1 & 1 & 0 \\
0 & 1 & 0 & 0 & 1 & 2 & 1 & 1 & 0 \\
\hline
0 & 1 & 1 & 0 & 1 & 0 & 1 & 0 & 1 \\
0 & 1 & 2 & 0 & 1 & 0 & 1 & 0 & 1 \\
0 & 2 & 1 & 0 & 0 & 1 & 1 & 0 & 1 \\
0 & 0 & 0 & 0 & 2 & 1 & 1 & 0 & 1 \\
0 & 0 & 1 & 0 & 2 & 1 & 1 & 0 & 1 \\
0 & 1 & 0 & 0 & 1 & 2 & 1 & 0 & 1
\end{array}
\right]
\right)
$$

$$
S_{h_2} = \left(
\begin{bmatrix}
112.980 \\
-26.800 \\
-13.400 \\
-2.680 \\
-2.680 \\
13.400 \\
2.680 \\
2.680 \\
\hline
-107.200 \\
\hline
-53.600 \\
\hline
-1477.386 \\
-131.454 \\
-131.454
\end{bmatrix}
,\;
\left[
\begin{array}{cccccc|ccc}
0 & 1 & 0 & 0 & 1 & 0 & 2 & 0 & 0 \\
0 & 1 & 1 & 0 & 1 & 0 & 2 & 0 & 0 \\
0 & 2 & 0 & 0 & 0 & 1 & 2 & 0 & 0 \\
0 & 1 & 2 & 0 & 1 & 0 & 2 & 0 & 0 \\
0 & 2 & 1 & 0 & 0 & 1 & 2 & 0 & 0 \\
0 & 0 & 0 & 0 & 2 & 1 & 2 & 0 & 0 \\
0 & 0 & 1 & 0 & 2 & 1 & 2 & 0 & 0 \\
0 & 1 & 0 & 0 & 1 & 2 & 2 & 0 & 0 \\
\hline
0 & 0 & 0 & 0 & 0 & 1 & 0 & 1 & 1 \\
\hline
0 & 0 & 0 & 0 & 0 & 1 & 0 & 0 & 2 \\
\hline
0 & 0 & 0 & 0 & 1 & 0 & 0 & 0 & 0 \\
0 & 0 & 1 & 0 & 1 & 0 & 0 & 0 & 0 \\
0 & 1 & 0 & 0 & 0 & 1 & 0 & 0 & 0
\end{array}
\right]
\right)
$$

$$
S_{h_3} = \left(\begin{bmatrix} -13.400 \\ -2.680 \\ -2.680 \\ 13.400 \\ 2.680 \\ 2.680 \\ \hline 53.600 \\ \hline -131.454 \\ -131.454 \end{bmatrix}, \quad \left[\begin{array}{cccccc|ccc} 0 & 1 & 1 & 0 & 1 & 0 & 2 & 0 & 0 \\ 0 & 1 & 2 & 0 & 1 & 0 & 2 & 0 & 0 \\ 0 & 2 & 1 & 0 & 0 & 1 & 2 & 0 & 0 \\ 0 & 0 & 0 & 0 & 2 & 1 & 2 & 0 & 0 \\ 0 & 0 & 1 & 0 & 2 & 1 & 2 & 0 & 0 \\ 0 & 1 & 0 & 0 & 1 & 2 & 2 & 0 & 0 \\ \hline 0 & 0 & 0 & 0 & 0 & 1 & 0 & 2 & 0 \\ \hline 0 & 0 & 1 & 0 & 1 & 0 & 0 & 0 & 0 \\ 0 & 1 & 0 & 0 & 0 & 1 & 0 & 0 & 0 \end{array} \right] \right)
$$

To these polynomial matrices we may now apply the approximation procedure from Para. 3.7. Since the admissible deviation is 5%, the elimination of rows may be performed according to Table 5.20. The rows being eliminated are not unifoldly determined, so the data from Table 5.20 represent only one of the possible combinations of rows which can be eliminated. As may be seen, about 30% of the total number of rows of the polynomial matrices considered has been eliminated. It should be noted, however, that this does not mean that the number of multiplications/additions is also reduced by 30%, since such a procedure degrades the property of antisymmetry of C – matrices which are incorporated into polynomial matrices S_{h_i}, $i \in N$. For example, by eliminating rows 2÷8 from

i	Maximal coefficient V_i^{max}	5% V_i^{max}	Eliminated rows
1	451.92	22.596	13.14
2	1477.386	73.869	2 ÷ 8
3	131.454	6.573	5.6

Table 5.20. Procedure for simplification of structural matrices S_{h_i} of arthropoid robot

S_{h_2}, the components of polynomial matrix $S_{C_{21}^1}$ (except for the first row) which also figures in S_{h_1} are eliminated. Accordingly, elimination of rows 2÷8 from S_{h_2} does not result in a reduced number of multiplications/additions. On the other hand, by eliminating rows 13 and 14 from S_{h_1}, and 5 and 6 from S_{h_3}, the number of operations does reduce. This leads to the conclusion that, after simplifying the model matrices, one must see the amount of savings in the number of operations, i.e., in computer time, and only on the basis of this decide whether or not it is justifiable to introduce the simplified model. Comparison of the number of operations for computation the exact and approximate model of arthropoid robot will be given in the text to follow.

Computational optimization of the analytical model of arthropoid robot and generation of program code

Let us now consider computation of the matrices of the dynamic model of arthropoid robot. The polynomial matrices of the dynamic model of this robot are given in Tables 5.15 and 5.16. According to the procedure described in Ch. 4, for each of these matrices it is necessary to generate the sequence of optimal calculation. The optimal sequences are described by two matrices (E and V) which determine the nodes and branches of the corresponding graphs. The results are summed in Tables 5.21, 5.22. and 5.23. In these tables we assumed

i,j	$E_{H_{ij}}$	$X_k,\ k>6$	$V_{H_{ij}}$
1,1	$\begin{bmatrix} 5 & 5 & 3 & 3 & 11 \\ 6 & 6 & 8 & 9 & 12 \\ 7 & 6 & 2 & 3 & 13 \\ 0 & 0 & 2 & 10 & 0 \\ 0 & 0 & 0 & 2 & 0 \end{bmatrix}$	$\begin{aligned} X_7 &= 4.57 \\ X_8 &= 112.98 \\ X_9 &= 26.80 \\ X_{10} &= 26.80 \\ X_{11} &= 2.68 \\ X_{12} &= 5.36 \\ X_{13} &= 2.68 \end{aligned}$	$\begin{bmatrix} 2 & 2 & 2 & 1 & 0 \\ 1 & 1 & 0 & 0 & 0 \\ 0 & 1 & 2 & 1 & 0 \\ 0 & 0 & 1 & 0 & 0 \\ 0 & 0 & 0 & 1 & 0 \end{bmatrix}$
2,2	$\begin{bmatrix} 3 & 8 \\ 7 & 0 \end{bmatrix}$	$\begin{aligned} X_7 &= 138.62 \\ X_8 &= 26.80 \end{aligned}$	$\begin{bmatrix} 1 & 0 \\ 0 & 0 \end{bmatrix}$
2,3	$\begin{bmatrix} 3 & 8 \\ 7 & 0 \end{bmatrix}$	$\begin{aligned} X_7 &= 3.90 \\ X_8 &= 13.40 \end{aligned}$	$\begin{bmatrix} 1 & 0 \\ 0 & 0 \end{bmatrix}$

Table 5.21. Matrix representatives of the elementary graphs of the inertial matrix of arthropoid robot

i	k,ℓ	$E\ {}_{C^i_{k\ell}}$					$x_k,\ k>6$	$V\ {}_{C^i_{k\ell}}$				
1	2,1	$\begin{bmatrix} 5 & 2 & 3 & 3 & 10 \\ 2 & 5 & 7 & 8 & 14 \\ 0 & 2 & 6 & 6 & 13 \\ 0 & 0 & 6 & 3 & 11 \\ 0 & 0 & 6 & 12 & 0 \\ 0 & 0 & 0 & 3 & 0 \\ 0 & 0 & 0 & 9 & 0 \end{bmatrix}$					$\begin{aligned} x_7 &= 112.98 \\ x_8 &= 26.80 \\ x_9 &= 13.40 \\ x_{10} &= 2.68 \\ x_{11} &= 2.68 \\ x_{12} &= -13.40 \\ x_{13} &= -2.68 \\ x_{14} &= -2.68 \end{aligned}$	$\begin{bmatrix} 2 & 3 & 2 & 1 & 0 \\ 1 & 1 & 0 & 0 & 0 \\ 0 & 1 & 1 & 1 & 0 \\ 0 & 0 & 2 & 0 & 0 \\ 0 & 0 & 2 & 1 & 0 \\ 0 & 0 & 0 & 1 & 0 \\ 0 & 0 & 0 & 0 & 0 \end{bmatrix}$				
1	3,1	$\begin{bmatrix} 5 & 2 & 3 & 3 & 8 \\ 2 & 5 & 6 & 7 & 12 \\ 0 & 2 & 6 & 6 & 11 \\ 0 & 0 & 3 & 3 & 9 \\ 0 & 0 & 0 & 10 \\ 0 & 0 & 0 & 6 \end{bmatrix}$					$\begin{aligned} x_7 &= 13.40 \\ x_8 &= 2.68 \\ x_9 &= 2.68 \\ x_{10} &= -13.40 \\ x_{11} &= -2.68 \\ x_{12} &= -2.68 \end{aligned}$	$\begin{bmatrix} 2 & 2 & 2 & 1 & 0 \\ 1 & 1 & 1 & 0 & 0 \\ 0 & 1 & 2 & 1 & 0 \\ 0 & 0 & 1 & 1 & 0 \\ 0 & 0 & 0 & 0 & 0 \\ 0 & 0 & 0 & 1 & 0 \end{bmatrix}$				
2	3,2	$[6 \quad 7]$					$x_7 = -13.40$	$[1 \quad 0]$				
2	3,3	$[6 \quad 7]$					$x_7 = -13.40$	$[1 \quad 0]$				

Table 5.22. Matrix representatives of the elementary graphs of
C - matrices of arthropoid robot

that $x_i = \cos q_i$, $x_{3+i} = \sin q_i$, and $i \in 1,2,3$ and that x_k, $k>6$ are constants.

A procedure for forming the program code on the basis of data from the
previous three tables was presented in Para. 4.3. The program code obtained may be used for real-time calculation of the matrices of dynamic
model. It should be noted that this program code is automatically obtained, i.e. by another program - the so-called expert-program (developed on the basis of the algorithm from Para. 4.3). In the case of FORTRAN,
the expert-program automatically generates the instructions of SUBROUTINE,
COMMON, RETURN and END type as well. The program code is thus completely prepared for compilation. Listing of the generated program code

i	X_k, $k > 6$		
2	$\begin{bmatrix} 5 & 3 & 8 \\ 2 & 7 & 9 \\ 0 & 6 & 0 \end{bmatrix}$	$X_7 = -1477.386$ $X_8 = -131.454$ $X_9 = -131.454$	$\begin{bmatrix} 2 & 1 & 0 \\ 1 & 0 & 0 \\ 0 & 1 & 0 \end{bmatrix}$
3	$\begin{bmatrix} 5 & 3 & 7 \\ 6 & 2 & 8 \end{bmatrix}$	$X_7 = -131.454$ $X_8 = -131.454$	$\begin{bmatrix} 1 & 1 & 0 \\ 1 & 1 & 0 \end{bmatrix}$

Table 5.23. Matrix representatives of the elementary graphs
of the gravity vector of arthropoid robot

is given in Fig. 5.8. Let us also state that the inputs to subroutines
are the internal angles and velocities, which are entered through COMMON
/UG/. Outputs of subroutines are the matrices of dynamic model

$$H(q, \theta) \quad \text{and} \quad h(q, \dot{q}, \theta)$$

where $h = \dot{q}^T C \dot{q} + h^G$. These data are in the COMMON /MATRIC/.

The number of operations required to form the model of arthropoid robot, according to ARTROP program shown in Fig. 5.8, is summed in Table
5.24. One should have in mind that the operations of COS/SIN type may

COS/SIN	Floating-point multiplications	Floating-point additions/subtractions
4	49	23

Table 5.24. The number of numerical operations for forming
the matrices of the model of arthropoid robot

be realized using table look-up and linear interpolation, which results
in a slight increase in the number of multiplications and additions. Of
course, accuracy of calculation then usually ranges from 2 to 4 significant digits, and this is often sufficient.

In this case the time to form the model using PDP 11/70 computer is

$$t \simeq 500 \ \mu s,$$

```
      SUBROUTINE ARTROP
C
      COMMON/MATRIC/H(6,6),C(6,6,6),G(6)
      COMMON/UG/Q(6)
C
      X7   = Q (1   )
      X8   = Q (2   )
      X2   = C O S ( X8   )
      X5   = S I N ( X8   )
      X9   = Q (3   )
      X3   = C O S ( X9   )
      X6   = S I N ( X9   )
C
      P2=X3*(-0.13145E+03)
      P1=X2*(-0.13145E+03)
      Q2=X5*P2
      Q1=X6*P1
      G(3)=Q1+Q2
      P2=X2*(-0.13145E+03)
      P1=X3*(-0.13145E+03)
      Q2=X6*P2
      Q1=X5*(P1  -0.14774E+04)
      G(2)=Q1+Q2
      G(1)=0.
      Q1=X6* 0.13400E+02
      C(3,2,2)=Q1
      C(3,2,1)=0.
      P4=X2*(-0.26800E+01)
      P3=X3*(-0.26800E+01)
      P2=X2* 0.26800E+01
      P1=X3* 0.26800E+01
      Q4=X2*P4
      Q3=X2*(P3  -0.13400E+02)
      Q2=X6*P2
      Q1=X5*(P1+ 0.13400E+02)
      P3=X3*Q4
      P2=X3*Q3
      P1=X6*(Q1+Q2)
      Q2=X6*P3
      Q1=X5*(P1+P2)
      C(3,1,1)=Q1+Q2
      Q1=X6*(-0.13400E+02)
      C(2,3,3)=Q1
      P4=X3*(-0.26800E+01)
      P3=X3* 0.26800E+01
      P2=X6* 0.26800E+01
      P1=X3*(-0.26800E+01)
      Q4=X2*(P4  -0.13400E+02)
      Q3=X5*(P3+ 0.13400E+02)
      Q2=X6*P2
      Q1=X3*(P1  -0.26800E+02)
      P3=X2*Q4
      P2=X6*Q3
      P1=X2*(Q1+Q2  -0.11298E+03)
      Q2=X6*P3
      Q1=X5*(P1+P2)
      C(2,1,1)=Q1+Q2
      C(1,3,3)=0.
```

```
                    C(1,3,2)=0.
                    C(1,2,2)=0.
                    H(3,3)=        0.39000E+01
                    Q1=X3*  0.13400E+02
                    H(3,2)=Q1+  0.39000E+01
                    H(3,1)=0.
                    Q1=X3*  0.26800E+02
                    H(2,2)=Q1+  0.13862E+03
                    H(2,1)=0.
                    P3=X2*  0.26800E+01
                    P2=X3*  0.53600E+01
                    P1=X3*  0.26800E+01
                    Q3=X2*P3
                    Q2=X2*(P2+  0.26800E+02)
                    Q1=X3*(P1+  0.26800E+02)
                    P3=X6*Q3
                    P2=X6*Q2
                    P1=X5*(Q1+  0.11298E+03)
                    Q2=X6*P3
                    Q1=X5*(P1+P2)
                    H(1,1)=Q1+Q2+  0.45700E+01
          C
                    RETURN
                    END
```

Fig. 5.8. Program for computing dynamic model matrices of
          arthropoid robot, obtained as output of expert-program

and in the case of calculating SIN/COS with 6 significant digits is
about 1 ms.

In generating the program code of approximate model (see Table 5.20),
the number of operations reduces by about 10 multiplications, which
represents savings of about

$$\frac{-\Delta n_{mul}}{n_{mul}} = 17\%$$

With robots having a larger number of degrees of freedom, considerably
higher savings in the number of operations are achievable.

5.4. An Anthropomorphic Robot

Let us consider a robot of anthropomorphic structure, shown in Fig.
5.9.

The model of anthropomorphic robot is shown in Fig. 5.10. The links
are denoted by c_1, c_2 and c_3, and joint connections by z_1, z_2 and z_3.
Like the arthropoid, this robot also has 3 rotational degrees of fre-

edom. They differ in that axes $\vec{e}_2$ and $\vec{e}_3$ with the arthropoid robot are parallel, and orthogonal with the anthropomorphic. This is why the mathematical model of this robot is more complex than in the preceding two examples.

Fig. 5.9. Anthropomorphic robot

The links with all geometrical parameters are separately presented in Fig. 5.11. Data on the kinematic parameters of links $\tilde{R}_i = \left\{\tilde{r}_{ii}, \ \tilde{r}_{i,i+1}\right\}$ and $\tilde{E}_i = \{\tilde{e}_i\}$ for $i \in N = \{1,2,3\}$ are given in Table 5.25.

i	$\tilde{r}_{ii}$ [m]	$\tilde{r}_{i,i+1}$ [m]	$\tilde{e}_i$	ξ_i
1	(0., 0., 0.50)	(0.14, 0., 0.)	(0., 0., 1.)	0
2	(0.175, 0., 0.)	(-0.175, 0., 0.)	(0., 1., 0.)	0
3	(0.230, 0., 0.)	(-0.130, 0., 0.)	(0., 0., 1.)	0

Table 5.25. Kinematic parameters of anthropomorphic robot

The dynamic parameters are given in Table 5.26. The parameters which are not included in this table are not important for constructing the dynamic model of anthropomorphic robot.

i	m_i [kg]	J_{i1} [kg m^2]	J_{i2} [kg m^2]	J_{i3} [kg m^2]
1	–	–	–	0.16
2	2.450	0.00377	0.03530	0.03530
3	0.723	0.00031	0.00488	0.00488

Table 5.26. Dynamic parameters of anthropomorphic robot

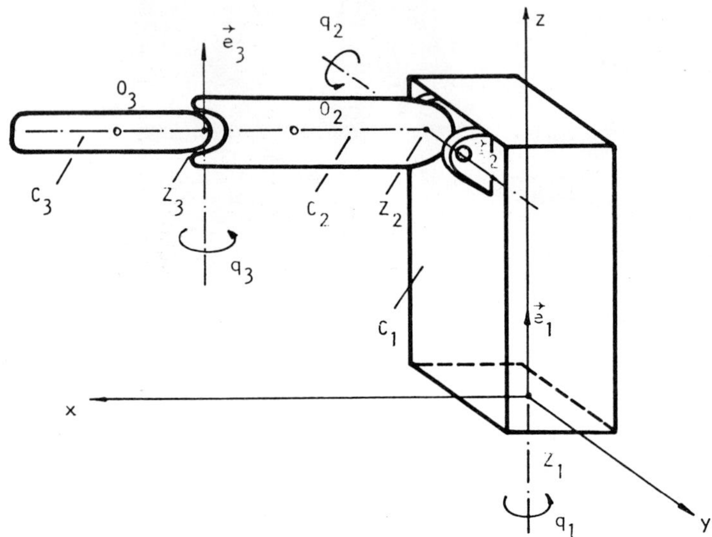

Fig. 5.10. Model of the anthropomorphic manipulator

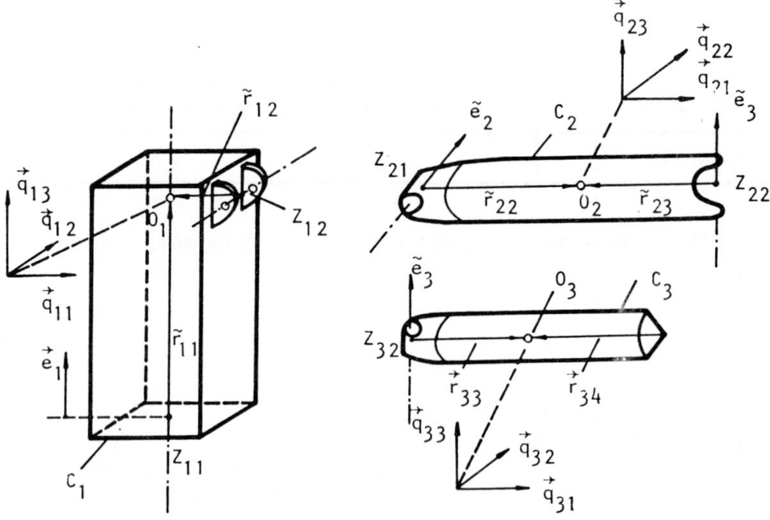

Fig.5.11. Kinematic parameters of links of anthropomorphic manipulator

Construction of nonlinear symbolic model of anthropomorphic robot

At the first stage of method for forming the nonlinear analytical model (Para. 2.3, Fig. 2.7) the "home" position of the mechanism is determined. The output of this stage are the matrices of transformation from local into reference coordinate system in the form

$$Q_i^O = \begin{bmatrix} 1 & 0 & 0 \\ 0 & 1 & 0 \\ 0 & 0 & 1 \end{bmatrix}, \qquad (i=1,2,3),$$

$$\vec{r}_{11}^{\,O} = \tilde{r}_{ii}, \qquad \vec{r}_{i,i+1}^{\,O} = \tilde{r}_{i,i+1}, \qquad (i=1,2,3),$$

and

$$\vec{e}_{i+1}^{\,O} = \tilde{e}_{i+1}, \qquad (i=1,2)$$

In the next stage of the algorithm, the position of mechanism links as a function of joint coordinates q_1, q_2 and q_3 is determined. The output of this stage are the polynomial matrices (see Ch. 2.5)

$$S_{q_{ij}}, \quad S_{r_{ii}}, \quad S_{r_{i,i+1}}, \quad S_{e_i}, \qquad i \in N, \qquad j=1,2,3.$$

Data given in Tables 5.27. and 5.28. are obtained by computer-aided generation of these matrices. The matrices of exponents have 6 columns, because the 7th, 8th and 9th columns are zero vectors. This is a result of the absence of prismatic degrees of freedom.

i	j	$S_{q_{ij}}$
1	1	$\left(\begin{bmatrix} 1. & 0. & 0. \\ 0. & 1. & 0. \end{bmatrix}, \begin{bmatrix} 1 & 0 & 0 & 0 & 0 & 0 \\ 0 & 0 & 0 & 1 & 0 & 0 \end{bmatrix} \right)$
1	2	$\left(\begin{bmatrix} 0. & 1. & 0. \\ -1. & 0. & 0. \end{bmatrix} \begin{bmatrix} 1 & 0 & 0 & 0 & 0 & 0 \\ 0 & 0 & 0 & 1 & 0 & 0 \end{bmatrix} \right)$
	3	$([0. \quad 0. \quad 1.], \quad [0\ 0\ 0 \quad 0\ 0\ 0])$

Table 5.27. Polynomial matrices corresponding to the unit vectors of the local coordinate systems of anthropomorphic robot

2	1	$\left(\begin{bmatrix} 1. & 0. & 0. \\ 0. & 1. & 0. \\ 0. & 0. & -1. \end{bmatrix} , \begin{bmatrix} 1 & 1 & 0 & 0 & 0 & 0 \\ 0 & 1 & 0 & 1 & 0 & 0 \\ 0 & 0 & 0 & 0 & 1 & 0 \end{bmatrix} \right)$
	2	$\left(\begin{bmatrix} 0. & 1. & 0. \\ -1. & 0. & 0. \end{bmatrix} , \begin{bmatrix} 1 & 0 & 0 & 0 & 0 & 0 \\ 0 & 0 & 0 & 1 & 0 & 0 \end{bmatrix} \right)$
	3	$\left(\begin{bmatrix} 0. & 0. & 1. \\ 1. & 0. & 0. \\ 0. & 1. & 0. \end{bmatrix} , \begin{bmatrix} 0 & 1 & 0 & 0 & 0 & 0 \\ 1 & 0 & 0 & 0 & 1 & 0 \\ 0 & 0 & 0 & 1 & 1 & 0 \end{bmatrix} \right)$
3	1	$\left(\begin{bmatrix} 1. & 0. & 0. \\ 0. & 1. & 0. \\ 0. & 0. & -1. \\ 0. & 1. & 0. \\ -1. & 0. & 0. \end{bmatrix} , \begin{bmatrix} 1 & 1 & 1 & 0 & 0 & 0 \\ 0 & 1 & 1 & 1 & 0 & 0 \\ 0 & 0 & 1 & 0 & 1 & 0 \\ 1 & 0 & 0 & 0 & 0 & 1 \\ 0 & 0 & 0 & 1 & 0 & 1 \end{bmatrix} \right)$
	2	$\left(\begin{bmatrix} 0. & 1. & 0. \\ -1. & 0. & 0. \\ -1. & 0. & 0. \\ 0. & 0. & 1. \\ 0. & -1. & 0. \end{bmatrix} , \begin{bmatrix} 1 & 0 & 1 & 0 & 0 & 0 \\ 0 & 0 & 1 & 1 & 0 & 0 \\ 1 & 1 & 0 & 0 & 0 & 1 \\ 0 & 0 & 0 & 0 & 1 & 1 \\ 0 & 1 & 0 & 1 & 0 & 1 \end{bmatrix} \right)$
	3	$\left(\begin{bmatrix} 0. & 0. & 1. \\ 1. & 0. & 0. \\ 0. & 1. & 0. \end{bmatrix} , \begin{bmatrix} 0 & 1 & 0 & 0 & 0 & 0 \\ 1 & 0 & 0 & 0 & 1 & 0 \\ 0 & 0 & 0 & 1 & 1 & 0 \end{bmatrix} \right)$

Table 5.27. (Contd.)

Polynomial matrices of the elements of the inertial matrix are formed in the next stage of the algorithm. The results are given in Table 5.29. Data from the upper triangular matrix are given only, because of the symmetry of the inertial matrix. It should be noted that the 2nd and 3rd degrees of freedom are inertially decoupled. Since $H_{33}=0.04305=$ =const., it follows that the 3rd joint is loaded by a constant inertial load.

i	S_{e_i}		
1	([0. 0. 1.] , [0 0 0 0 0 0])		
2	$\left(\begin{bmatrix} 0. & 1. & 0. \\ -1. & 0. & 0. \end{bmatrix} , \begin{bmatrix} 1 & 0 & 0 & 0 & 0 & 0 \\ 0 & 0 & 0 & 1 & 0 & 0 \end{bmatrix} \right)$		
3	$\left(\begin{bmatrix} 0. & 0. & 1. \\ 1. & 0. & 0. \\ 0. & 1. & 0. \end{bmatrix} , \begin{bmatrix} 0 & 1 & 0 & 0 & 0 & 0 \\ 1 & 0 & 0 & 0 & 1 & 0 \\ 0 & 0 & 0 & 1 & 1 & 0 \end{bmatrix} \right)$		

i	$S_{r_{ii}}$		
1	([0. 0. 0.5], [0 0 0 0 0 0])		
2	$\left(\begin{bmatrix} 0.175 & 0. & 0. \\ 0. & 0.175 & 0. \\ 0. & 0. & -0.175 \end{bmatrix} , \begin{bmatrix} 1 & 1 & 0 & 0 & 0 & 0 \\ 0 & 1 & 0 & 1 & 0 & 0 \\ 0 & 0 & 0 & 0 & 1 & 0 \end{bmatrix} \right)$		
3	$\left(\begin{bmatrix} 0.23 & 0. & 0. \\ 0. & 0.23 & 0. \\ 0. & 0. & -0.23 \\ 0. & 0.23 & 0. \\ -0.23 & 0. & 0. \end{bmatrix} , \begin{bmatrix} 1 & 1 & 1 & 0 & 0 & 0 \\ 0 & 1 & 1 & 1 & 0 & 0 \\ 0 & 0 & 1 & 0 & 1 & 0 \\ 1 & 0 & 0 & 0 & 0 & 1 \\ 0 & 0 & 0 & 1 & 0 & 1 \end{bmatrix} \right)$		

i	$S_{r_{i,i+1}}$		
1	$\left(\begin{bmatrix} -0.14 & 0. & 0. \\ 0. & -0.14 & 0. \end{bmatrix} , \begin{bmatrix} 1 & 0 & 0 & 0 & 0 & 0 \\ 0 & 0 & 0 & 1 & 0 & 0 \end{bmatrix} \right)$		
2	$\left(\begin{bmatrix} -0.175 & 0. & 0. \\ 0. & -0.175 & 0. \\ 0. & 0. & 0.175 \end{bmatrix} , \begin{bmatrix} 1 & 1 & 0 & 0 & 0 & 0 \\ 0 & 1 & 0 & 1 & 0 & 0 \\ 0 & 0 & 0 & 0 & 1 & 0 \end{bmatrix} \right)$		

Table 5.28. Polynomial matrices S_{e_i}, $S_{r_{ii}}$ and $S_{r_{i,i+1}}$ of anthropomorphic robot

The next stage of the algorithm consists of generating the polynomial matrices of the elements of C - matrices (Table 5.30). It may be seen that elements C_{11}^{1}, C_{21}^{2}, C_{22}^{2}, C_{33}^{2}, C_{31}^{3}, C_{32}^{3} and C_{33}^{3} are zero elements. The remaining elements, which are not given in Table 5.30, are directly obtained from the property of symmetry and antisymmetry of C - matrices (see Para 2.4).

i,j	$^{S}H_{ij}$
1,1	$\left(\begin{bmatrix} 0.22619 \\ 0.19520 \\ 0.19090 \\ 0.04275 \\ 0.04275 \\ 0.04656 \\ 0.11640 \end{bmatrix} , \begin{bmatrix} 0 & 0 & 0 & 0 & 0 & 0 \\ 0 & 2 & 0 & 0 & 0 & 0 \\ 0 & 1 & 0 & 0 & 0 & 0 \\ 0 & 2 & 2 & 0 & 0 & 0 \\ 0 & 0 & 0 & 0 & 0 & 2 \\ 0 & 1 & 1 & 0 & 0 & 0 \\ 0 & 2 & 1 & 0 & 0 & 0 \end{bmatrix} \right)$
1,2	$\left(\begin{bmatrix} 0.04275 \\ 0.05820 \end{bmatrix} , \begin{bmatrix} 0 & 0 & 1 & 0 & 1 & 1 \\ 0 & 0 & 0 & 0 & 1 & 1 \end{bmatrix} \right)$
1,3	$\left(\begin{bmatrix} 0.04305 \\ 0.02328 \\ 0.05820 \end{bmatrix} , \begin{bmatrix} 0 & 1 & 0 & 0 & 0 & 0 \\ 0 & 0 & 1 & 0 & 0 & 0 \\ 0 & 1 & 1 & 0 & 0 & 0 \end{bmatrix} \right)$
2,2	$\left(\begin{bmatrix} 0.19920 \\ 0.04275 \\ 0.11640 \end{bmatrix} , \begin{bmatrix} 0 & 0 & 0 & 0 & 0 & 0 \\ 0 & 0 & 2 & 0 & 0 & 0 \\ 0 & 0 & 1 & 0 & 0 & 0 \end{bmatrix} \right)$
2,3	$\left(\begin{bmatrix} 0.00000 \end{bmatrix} , \begin{bmatrix} 0 & 0 & 0 & 0 & 0 & 0 \end{bmatrix} \right)$
3,3	$\left(\begin{bmatrix} 0.04305 \end{bmatrix} , \begin{bmatrix} 0 & 0 & 0 & 0 & 0 & 0 \end{bmatrix} \right)$

Table 5.29. Polynomial matrices of the elements of the inertial matrix

The last stage of the algorithm includes calculation of the polynomial matrices of gravity vector $^{S}G_{h_{i}}$, $i \in N$. Thus, one obtains

i	k,ℓ	$S_{C_{k\ell}^i}$
1	1,1	$\left(\begin{bmatrix} -0.09545 \\ -0.19705 \\ -0.11640 \\ -0.02330 \\ -0.04290 \end{bmatrix}, \begin{bmatrix} 0 & 0 & 0 & 0 & 1 & 0 \\ 0 & 1 & 0 & 0 & 1 & 0 \\ 0 & 1 & 1 & 0 & 1 & 0 \\ 0 & 0 & 1 & 0 & 1 & 0 \\ 0 & 1 & 2 & 0 & 1 & 0 \end{bmatrix} \right)$
1	3,1	$\left(\begin{bmatrix} 0.03825 \\ -0.02330 \\ -0.05820 \\ -0.03825 \\ 0.00465 \end{bmatrix}, \begin{bmatrix} 0 & 0 & 1 & 0 & 0 & 1 \\ 0 & 1 & 0 & 0 & 0 & 1 \\ 0 & 2 & 0 & 0 & 0 & 1 \\ 0 & 2 & 1 & 0 & 0 & 1 \\ 0 & 0 & 1 & 0 & 2 & 1 \end{bmatrix} \right)$
1	2,2	$\left(\begin{bmatrix} 0.05820 \\ 0.04300 \end{bmatrix}, \begin{bmatrix} 0 & 1 & 0 & 0 & 0 & 1 \\ 0 & 1 & 1 & 0 & 0 & 1 \end{bmatrix} \right)$
1	3,2	$\left(\begin{bmatrix} -0.04050 \\ -0.00255 \\ 0.00240 \end{bmatrix}, \begin{bmatrix} 0 & 0 & 0 & 0 & 1 & 2 \\ 0 & 0 & 0 & 0 & 1 & 0 \\ 0 & 0 & 2 & 0 & 1 & 0 \end{bmatrix} \right)$
1	3,3	$\left(\begin{bmatrix} -0.02330 \\ -0.05820 \end{bmatrix}, \begin{bmatrix} 0 & 0 & 0 & 0 & 0 & 1 \\ 0 & 1 & 0 & 0 & 0 & 1 \end{bmatrix} \right)$
2	3,1	$\left(\begin{bmatrix} 0.05820 \\ 0.04050 \\ 0.00255 \\ -0.00240 \end{bmatrix}, \begin{bmatrix} 0 & 0 & 1 & 0 & 1 & 0 \\ 0 & 0 & 2 & 0 & 1 & 0 \\ 0 & 0 & 0 & 0 & 1 & 0 \\ 0 & 0 & 0 & 0 & 1 & 2 \end{bmatrix} \right)$
2	3,2	$\left(\begin{bmatrix} -0.05820 \\ -0.0428 \end{bmatrix}, \begin{bmatrix} 0 & 0 & 0 & 0 & 0 & 1 \\ 0 & 0 & 1 & 0 & 0 & 1 \end{bmatrix} \right)$

Table 5.30. Polynomial matrices of nonzero elements of C - matrices

$$S{}_{h^G_i} = 0$$

$$S{}_{h^G_2} = \left(\begin{bmatrix} -6.68850 \\ -1.63130 \end{bmatrix}, \begin{bmatrix} 0 & 1 & 0 & 0 & 0 & 0 \\ 0 & 1 & 1 & 0 & 0 & 0 \end{bmatrix} \right)$$

$$S{}_{h^G_3} = ([1.63130], [0 \; 0 \; 0 \quad 0 \; 1 \; 1]).$$

The polynomial matrices of all elements of dynamic model matrices of anthropomorphic robot have been formed in this way.

Construction of linearized model of anthropomorphic robot

As described in the previous examples, construction of the linearized model requires the following polynomial matrices to be determined

$$S{}_{H_{ij}}(s), \quad S{}_{C^i_{jk}}(s) \quad \text{and} \quad S{}_{h^G_i}(s)$$

for $i,j,k, s \in N$, where (s) denotes the partial derivative with respect to joint coordinate q_s. Application of the algorithm from Para. 3.6 gives the results presented in Tables 5.31, 5.32 and 5.33. The polynomial matrices of elements which are not given in these tables are directly obtained by applying the properties of symmetry and antisymmetry (Para. 2.4), or are zero elements. Construction of the matrices of linearized model has thus been completed.

Computational optimization of the analytical model of anthropomorphic robot and generation of program code

In contrast to the preceding two examples, we will here consider generation of the matrices of the linearized model of anthropomorphic robot as well. Polynomial matrices of the linearized model are given in Tables 5.31 - 5.33. According to the algorithm presented in Ch. 4, for these matrices it is necessary to generate the corresponding sequences of optimal calculation, i.e., the matrix representatives of minimal graphs (E and V matrices). For the elements corresponding to the partial derivatives of inertial matrix one obtains the results given in Table 5.34. On the basis of these data it is easy to draw the graphs of optimal calculation. For example, for the term

i,j	s	$_S H_{ij}(s)$	
1,1	2	$\begin{bmatrix} -0.3904 \\ -0.19090 \\ -0.08550 \\ -0.04656 \\ -0.23280 \end{bmatrix}$	$\begin{bmatrix} 0 & 1 & 0 & 0 & 1 & 0 \\ 0 & 0 & 0 & 0 & 1 & 0 \\ 0 & 1 & 2 & 0 & 1 & 0 \\ 0 & 0 & 1 & 0 & 1 & 0 \\ 0 & 1 & 1 & 0 & 1 & 0 \end{bmatrix}$
1,1	3	$\begin{bmatrix} -0.08550 \\ 0.08550 \\ -0.04656 \\ -0.11640 \end{bmatrix}$	$\begin{bmatrix} 0 & 2 & 1 & 0 & 0 & 1 \\ 0 & 0 & 1 & 0 & 0 & 1 \\ 0 & 1 & 0 & 0 & 0 & 1 \\ 0 & 2 & 0 & 0 & 0 & 1 \end{bmatrix}$
1,2	2	$\begin{bmatrix} 0.04275 \\ 0.05820 \end{bmatrix}$	$\begin{bmatrix} 0 & 1 & 1 & 0 & 0 & 1 \\ 0 & 1 & 0 & 0 & 0 & 1 \end{bmatrix}$
1,2	3	$\begin{bmatrix} 0.04275 \\ -0.08550 \\ 0.05820 \end{bmatrix}$	$\begin{bmatrix} 0 & 0 & 0 & 0 & 1 & 0 \\ 0 & 0 & 0 & 0 & 1 & 2 \\ 0 & 0 & 1 & 0 & 1 & 0 \end{bmatrix}$
1,3	2	$\begin{bmatrix} -0.04305 \\ -0.05820 \end{bmatrix}$	$\begin{bmatrix} 0 & 0 & 0 & 0 & 1 & 0 \\ 0 & 0 & 1 & 0 & 1 & 0 \end{bmatrix}$
1,3	3	$\begin{bmatrix} -0.02328 \\ -0.05820 \end{bmatrix}$	$\begin{bmatrix} 0 & 0 & 0 & 0 & 0 & 1 \\ 0 & 1 & 0 & 0 & 0 & 1 \end{bmatrix}$
2,2	3	$\begin{bmatrix} -0.08550 \\ -0.11640 \end{bmatrix}$	$\begin{bmatrix} 0 & 0 & 1 & 0 & 0 & 1 \\ 0 & 0 & 0 & 0 & 0 & 1 \end{bmatrix}$

Table 5.31. Polynomial matrices of the partial derivatives
of nonzero elements of the inertial matrix

i	k,ℓ	s	$^S\!C_{ik\ell}^{i(s)}$
1	2,1	2	$\left(\begin{bmatrix} -0.09545 \\ -0.19705 \\ 0.39410 \\ -0.11640 \\ 0.23280 \\ -0.02330 \\ -0.04290 \\ 0.08580 \end{bmatrix}, \begin{bmatrix} 0 & 1 & 0 & 0 & 0 & 0 \\ 0 & 0 & 0 & 0 & 0 & 0 \\ 0 & 0 & 0 & 0 & 2 & 0 \\ 0 & 0 & 1 & 0 & 0 & 0 \\ 0 & 0 & 1 & 0 & 2 & 0 \\ 0 & 1 & 1 & 0 & 0 & 0 \\ 0 & 0 & 2 & 0 & 0 & 0 \\ 0 & 0 & 1 & 0 & 2 & 0 \end{bmatrix} \right)$
1	2,1	3	$\left(\begin{bmatrix} 0.11640 \\ 0.02330 \\ 0.08580 \end{bmatrix}, \begin{bmatrix} 0 & 1 & 0 & 0 & 1 & 1 \\ 0 & 0 & 0 & 0 & 1 & 1 \\ 0 & 1 & 1 & 0 & 1 & 1 \end{bmatrix} \right)$
1	3,1	2	$\left(\begin{bmatrix} 0.02330 \\ 0.11640 \\ 0.08580 \end{bmatrix}, \begin{bmatrix} 0 & 0 & 0 & 0 & 1 & 1 \\ 0 & 1 & 0 & 0 & 1 & 1 \\ 0 & 1 & 1 & 0 & 1 & 1 \end{bmatrix} \right)$
1	3,1	3	$\left(\begin{bmatrix} 0.03825 \\ -0.07650 \\ -0.02330 \\ -0.05820 \\ -0.03825 \\ 0.07650 \\ 0.00465 \\ -0.00930 \end{bmatrix}, \begin{bmatrix} 0 & 0 & 0 & 0 & 0 & 0 \\ 0 & 0 & 0 & 0 & 0 & 2 \\ 0 & 1 & 1 & 0 & 0 & 0 \\ 0 & 2 & 1 & 0 & 0 & 0 \\ 0 & 2 & 0 & 0 & 0 & 0 \\ 0 & 2 & 0 & 0 & 0 & 2 \\ 0 & 0 & 0 & 0 & 2 & 0 \\ 0 & 0 & 0 & 0 & 2 & 2 \end{bmatrix} \right)$
1	2,2	2	$\left(\begin{bmatrix} -0.05820 \\ -0.04300 \end{bmatrix}, \begin{bmatrix} 0 & 0 & 0 & 0 & 1 & 1 \\ 0 & 0 & 1 & 0 & 1 & 1 \end{bmatrix} \right)$

Table 5.32. Polynomial matrices of the partial derivatives
of the elements of C - matrices

1	2,2	3	$\left(\begin{bmatrix} 0.05820 \\ 0.04300 \\ -0.08600 \end{bmatrix}, \begin{bmatrix} 0 & 1 & 1 & 0 & 0 & 0 \\ 0 & 1 & 0 & 0 & 0 & 0 \\ 0 & 1 & 0 & 0 & 0 & 2 \end{bmatrix}\right)$
1	3,2	2	$\left(\begin{bmatrix} -0.04050 \\ -0.00255 \\ 0.00240 \end{bmatrix}, \begin{bmatrix} 0 & 1 & 0 & 0 & 0 & 2 \\ 0 & 1 & 0 & 0 & 0 & 0 \\ 0 & 1 & 2 & 0 & 0 & 0 \end{bmatrix}\right)$
1	3,2	3	$([-0.08580], [0\ 0\ 1\ 0\ 1\ 1])$
1	3,3	2	$([\,0.05820], [0\ 0\ 0\ 0\ 1\ 1])$
1	3,3	3	$\left(\begin{bmatrix} -0.02330 \\ -0.05820 \end{bmatrix}, \begin{bmatrix} 0 & 0 & 1 & 0 & 0 & 0 \\ 0 & 1 & 1 & 0 & 0 & 0 \end{bmatrix}\right)$
2	3,1	2	$\left(\begin{bmatrix} 0.05820 \\ 0.04050 \\ 0.00255 \\ -0.00240 \end{bmatrix}, \begin{bmatrix} 0 & 1 & 1 & 0 & 0 & 0 \\ 0 & 1 & 2 & 0 & 0 & 0 \\ 0 & 1 & 0 & 0 & 0 & 0 \\ 0 & 1 & 0 & 0 & 0 & 2 \end{bmatrix}\right)$
2	3,1	3	$\left(\begin{bmatrix} -0.05820 \\ -0.08580 \end{bmatrix}, \begin{bmatrix} 0 & 0 & 0 & 0 & 1 & 1 \\ 0 & 0 & 1 & 0 & 1 & 1 \end{bmatrix}\right)$
2	3,2	2	$([\,0.00000], [0\ 0\ 0\ 0\ 0\ 0])$
2	3,2	3	$\left(\begin{bmatrix} -0.05820 \\ -0.04820 \\ 0.09640 \end{bmatrix}, \begin{bmatrix} 0 & 0 & 1 & 0 & 0 & 0 \\ 0 & 0 & 0 & 0 & 0 & 0 \\ 0 & 0 & 0 & 0 & 0 & 2 \end{bmatrix}\right)$

Table 5.32. (Contd.)

$$H_{12}^{(3)} = \frac{\partial H_{12}(q,\ \theta)}{\partial q_3}$$

one obtains the graph

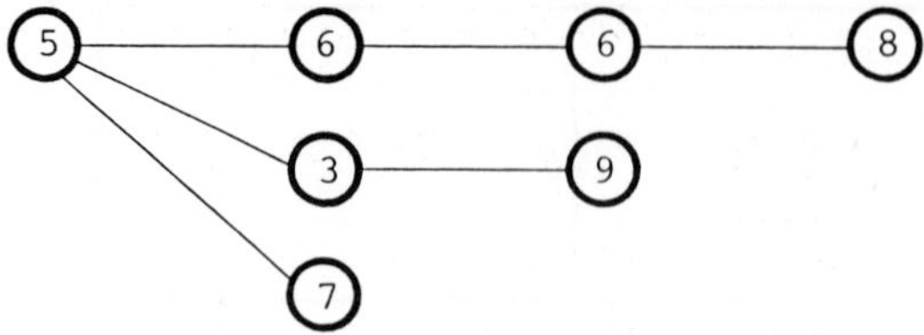

which can be interpreted by the following polynomial

$$H_{12}^{(3)} = x_5\left(x_6^2 x_8 + x_3 x_9 + x_7\right)$$

i.e., by trigonometric form

$$H_{12}^{(3)} = \sin q_2\left(\left(\sin^2 q_3\right)x_8 + \left(\cos q_3\right)x_9 + x_7\right)$$

where X_7, X_8 and X_9 are the constants given in Table 5.34.

i	s	$h_i^{G(s)}$	
2	2	$\left(\begin{bmatrix} 6.68850 \\ 1.63130 \end{bmatrix}\right.$, $\left.\begin{bmatrix} 0\ 0\ 0 & 0\ 1\ 0 \\ 0\ 0\ 1 & 0\ 1\ 0 \end{bmatrix}\right)$	
2	3	([1.63130] ,	[0 1 0 0 0 1])
3	2	([1.63130] ,	[0 1 0 0 0 1])
3	3	([1.63130] ,	[0 0 1 0 1 0])

Table 5.33. Polynomial matrices of the partial derivatives of
the elements of gravity vector

Matrices E and V, corresponding to the partial derivatives of C - matrices and gravity vector, are given in Tables 5.35. and 5.36, respectively. While, in the preceding examples, the program code was generated for the nonlinear dynamic model of the mechanism, it will be shown
here that it is possible to generate the program code for the linearized model in a practically identical manner. It should be noted that
this program code is also automatically obtained, by the so-called expert-program (see Para. 4.3). The expert-program also generates input/

i,j	s	$E_{H_{ij}}(s)$	X_k, k>6	$V_{H_{ij}}(s)$
1,1	2	$\begin{bmatrix} 5 & 2 & 3 & 3 & 9 \\ 0 & 3 & 7 & 11 & 0 \\ 0 & 8 & 10 & 0 & 0 \\ 0 & 0 & 0 & 0 & 0 \end{bmatrix}$	$X_7 = -0.3904$ $X_8 = -0.1909$ $X_9 = -0.08550$ $X_{10} = -0.04656$ $X_{11} = -0.23280$	$\begin{bmatrix} 3 & 2 & 2 & 1 & 0 \\ 0 & 1 & 0 & 0 & 0 \\ 0 & 0 & 0 & 0 & 0 \\ 0 & 0 & 0 & 0 & 0 \end{bmatrix}$
1,1	3	$\begin{bmatrix} 6 & 2 & 2 & 3 & 7 \\ 0 & 3 & 9 & 10 & 0 \\ 0 & 0 & 8 & 0 & 0 \end{bmatrix}$	$X_7 = -0.0855$ $X_8 = -0.0855$ $X_9 = -0.04656$ $X_{10} = -0.11640$	$\begin{bmatrix} 2 & 2 & 2 & 1 & 0 \\ 0 & 1 & 0 & 0 & 0 \\ 0 & 0 & 0 & 0 & 0 \end{bmatrix}$
1,2	2	$\begin{bmatrix} 2 & 6 & 3 & 7 \\ 0 & 0 & 8 & 0 \end{bmatrix}$	$X_7 = 0.04275$ $X_8 = 0.05820$	$\begin{bmatrix} 1 & 2 & 1 & 0 \\ 0 & 0 & 0 & 0 \end{bmatrix}$
1,2	3	$\begin{bmatrix} 5 & 6 & 6 & 8 \\ 0 & 3 & 9 & 0 \\ 0 & 7 & 0 & 0 \end{bmatrix}$	$X_7 = 0.04275$ $X_8 = -0.08550$ $X_9 = 0.05820$	$\begin{bmatrix} 3 & 1 & 1 & 0 \\ 0 & 1 & 0 & 0 \\ 0 & 0 & 0 & 0 \end{bmatrix}$
1,3	2	$\begin{bmatrix} 5 & 3 & 8 \\ 0 & 7 & 0 \end{bmatrix}$	$X_7 = -8.04305$ $X_8 = -0.05820$	$\begin{bmatrix} 2 & 1 & 0 \\ 0 & 0 & 0 \end{bmatrix}$
1,3	3	$\begin{bmatrix} 6 & 2 & 8 \\ 0 & 7 & 0 \end{bmatrix}$	$X_7 = -0.02328$ $X_8 = -0.05820$	$\begin{bmatrix} 2 & 1 & 0 \\ 0 & 0 & 0 \end{bmatrix}$
2,2	3	$\begin{bmatrix} 6 & 3 & 7 \\ 0 & 8 & 0 \end{bmatrix}$	$X_7 = -0.08550$ $X_8 = -0.11640$	$\begin{bmatrix} 2 & 1 & 0 \\ 0 & 0 & 0 \end{bmatrix}$

Table 5.34. Matrix representatives of the elementary graphs of the partial derivatives of the inertial matrix

/output instructions in a desired programming language. For example, instructions of SUBROUTINE, COMMON, etc. are generated in FORTRAN as well. The listing of the obtained program is given in Fig. 5.12. The

i	k,ℓ	s	$E_{C^{i(s)}_{k\ell}}$	$X_k,\ k>6$	$V_{C^{i(s)}_{k\ell}}$
1	2,1	2	$\begin{bmatrix} 3 & 5 & 5 & 3 & 14 \\ 2 & 2 & 12 & 11 & 0 \\ 5 & 3 & 13 & 0 & 0 \\ 8 & 10 & 9 & 0 & 0 \\ 0 & 7 & 0 & 0 & 0 \\ 0 & 5 & 0 & 0 & 0 \end{bmatrix}$	$\begin{aligned} X_7 &= -0.09545 \\ X_8 &= -0.19705 \\ X_9 &= 0.39410 \\ X_{10} &= -0.11640 \\ X_{11} &= 0.23280 \\ X_{12} &= -0.02330 \\ X_{13} &= -0.04290 \\ X_{14} &= 0.08580 \end{aligned}$	$\begin{bmatrix} 4 & 1 & 2 & 1 & 0 \\ 1 & 1 & 0 & 0 & 0 \\ 1 & 1 & 0 & 0 & 0 \\ 0 & 0 & 0 & 0 & 0 \\ 0 & 0 & 0 & 0 & 0 \\ 0 & 1 & 0 & 0 & 0 \end{bmatrix}$
1	2,1	3	$\begin{bmatrix} 5 & 6 & 2 & 3 & 9 \\ 0 & 0 & 8 & 7 & 0 \end{bmatrix}$	$\begin{aligned} X_7 &= 0.11640 \\ X_8 &= 0.02330 \\ X_9 &= 0.08580 \end{aligned}$	$\begin{bmatrix} 1 & 2 & 2 & 1 & 0 \\ 0 & 0 & 0 & 0 & 0 \end{bmatrix}$
1	3,1	2	$\begin{bmatrix} 5 & 6 & 2 & 3 & 9 \\ 0 & 0 & 7 & 8 & 0 \end{bmatrix}$	$\begin{aligned} X_7 &= 0.02330 \\ X_8 &= 0.11640 \\ X_9 &= 0.08580 \end{aligned}$	$\begin{bmatrix} 1 & 2 & 2 & 1 & 0 \\ 0 & 0 & 0 & 0 & 0 \end{bmatrix}$
1	3,1	3	$\begin{bmatrix} 2 & 2 & 6 & 6 & 12 \\ 5 & 3 & 3 & 10 & 14 \\ 6 & 5 & 11 & 6 & 0 \\ 7 & 6 & 9 & 0 & 0 \\ 0 & 0 & 6 & 0 & 0 \\ 0 & 0 & 13 & 0 & 0 \\ 0 & 0 & 8 & 0 & 0 \end{bmatrix}$	$\begin{aligned} X_7 &= 0.03825 \\ X_8 &= -0.07650 \\ X_9 &= -0.02330 \\ X_{10} &= -0.05820 \\ X_{11} &= -0.03825 \\ X_{12} &= 0.07650 \\ X_{13} &= 0.00465 \\ X_{14} &= -0.00930 \end{aligned}$	$\begin{bmatrix} 2 & 3 & 1 & 1 & 0 \\ 1 & 1 & 1 & 0 & 0 \\ 1 & 1 & 0 & 1 & 0 \\ 0 & 1 & 0 & 0 & 0 \\ 0 & 0 & 1 & 0 & 0 \\ 0 & 0 & 0 & 0 & 0 \\ 0 & 0 & 0 & 0 & 0 \end{bmatrix}$
1	2,2	2	$\begin{bmatrix} 5 & 6 & 3 & 8 \\ 0 & 0 & 7 & 0 \end{bmatrix}$	$\begin{aligned} X_7 &= -0.05820 \\ X_8 &= -0.04300 \end{aligned}$	$\begin{bmatrix} 1 & 2 & 1 & 0 \\ 0 & 0 & 0 & 0 \end{bmatrix}$

Table 5.35. Matrix representatives of elementary graphs of partial derivatives of C - matrices

1	2,2	3	$\begin{bmatrix} 2 & 6 & 6 & 9 \\ 0 & 3 & 7 & 0 \\ 0 & 8 & 0 & 0 \end{bmatrix}$	$\begin{aligned} X_7 &= 0.05820 \\ X_8 &= 0.04300 \\ X &= -0.08600 \end{aligned}$	$\begin{bmatrix} 3 & 1 & 1 & 0 \\ 0 & 1 & 0 & 0 \\ 0 & 0 & 0 & 0 \end{bmatrix}$
1	3,2	2	$\begin{bmatrix} 2 & 6 & 6 & 7 \\ 0 & 3 & 3 & 9 \\ 0 & 8 & 0 & 0 \end{bmatrix}$	$\begin{aligned} X_7 &= -0.04050 \\ X_8 &= -0.00255 \\ X_9 &= 0.00240 \end{aligned}$	$\begin{bmatrix} 3 & 1 & 1 & 0 \\ 0 & 1 & 1 & 0 \\ 0 & 0 & 0 & 0 \end{bmatrix}$
1	3,2	3	$\begin{bmatrix} 3 & 5 & 6 & 7 \end{bmatrix}$	$X_7 = -0.08580$	$\begin{bmatrix} 1 & 1 & 1 & 0 \end{bmatrix}$
1	3,3	2	$\begin{bmatrix} 5 & 6 & 7 \end{bmatrix}$	$X_7 = 0.05820$	$\begin{bmatrix} 1 & 1 & 0 \end{bmatrix}$
1	3,3	3	$\begin{bmatrix} 3 & 2 & 8 \\ 0 & 7 & 0 \end{bmatrix}$	$\begin{aligned} X_7 &= -0.02330 \\ X_8 &= -0.05820 \end{aligned}$	$\begin{bmatrix} 2 & 1 & 0 \\ 0 & 0 & 0 \end{bmatrix}$
2	3,1	2	$\begin{bmatrix} 2 & 3 & 3 & 8 \\ 0 & 6 & 7 & 10 \\ 0 & 9 & 6 & 0 \end{bmatrix}$	$\begin{aligned} X_7 &= 0.05820 \\ X_8 &= 0.04050 \\ X_9 &= 0.00255 \\ X_{10} &= -0.00240 \end{aligned}$	$\begin{bmatrix} 3 & 2 & 1 & 0 \\ 0 & 1 & 0 & 0 \\ 0 & 0 & 1 & 0 \end{bmatrix}$
2	3,1	3	$\begin{bmatrix} 5 & 6 & 3 & 8 \\ 0 & 0 & 7 & 0 \end{bmatrix}$	$\begin{aligned} X_7 &= -0.05820 \\ X_8 &= -0.08580 \end{aligned}$	$\begin{bmatrix} 1 & 2 & 1 & 0 \\ 0 & 0 & 0 & 0 \end{bmatrix}$
2	3,2	3	$\begin{bmatrix} 6 & 6 & 9 \\ 3 & 7 & 0 \\ 8 & 0 & 0 \end{bmatrix}$	$\begin{aligned} X_7 &= -0.05820 \\ X_8 &= -0.04820 \\ X_9 &= 0.09640 \end{aligned}$	$\begin{bmatrix} 1 & 1 & 0 \\ 1 & 0 & 0 \\ 0 & 0 & 0 \end{bmatrix}$

Table 5.35. (Contd.)

subroutine inputs are the internal angles $\left(q_1, \ldots, q_n\right)$, which are in the COMMON - block /UG/. The outputs of subroutine are the matrices of linearized model

$$\frac{\partial H(q, \theta)}{\partial q}, \quad \frac{\partial C^i(q, \theta)}{\partial q}, \quad \frac{\partial h^G(q, \theta)}{\partial q},$$

for $i \in N$. The elements of these matrices are stored in the COMMON-block

216

/LMODL/ in the following way

$$H_{ij}^{(s)} \rightarrow HL(i, j, s),$$

$$C_{k\ell}^{i(s)} \rightarrow CL(i, k, \ell, s)$$

$$h_i^{G(s)} \rightarrow HGL(i, s).$$

i	s	$E_{h_i^{G(s)}}$			$X_k,\ k>6$	$V_{h_i^{G(s)}}$		
2	2	$\begin{matrix}5 \\ 0\end{matrix}$	$\begin{matrix}3 \\ 7\end{matrix}$	$\begin{matrix}8 \\ 0\end{matrix}$	$X_7 = 6.68850$ $X_8 = 1.63130$	$\begin{matrix}2 \\ 0\end{matrix}$	$\begin{matrix}1 \\ 0\end{matrix}$	$\begin{matrix}0 \\ 0\end{matrix}$
2	3	[2	6	7]	$X_7 = 1.63130$	[1	1	0]
3	2	[2	6	7]	$X_7 = 1.63130$	[1	1	0]
3	3	[3	5	7]	$X_7 = 1.63130$	[1	1	0]

Table 5.36. Matrix representatives of elementary graphs of
partial derivatives of gravity vector

The program does not set zero elements nor the elements which directly
follow from the property of symmetry and antisymmetry of model matrices.
Table 5.37 shows the number of operations required to form the linearized model of anthropomorphic robot, according to LANTR program given
in Fig. 5.12. If cosines and sines are computed by table look-up method, the following time is required to form the model on PDP 11/70

$$t \approx 700\ \mu s$$

If functions SIN and COS are called from the program library (about 6
significant digits), the time to form the matrices of partial derivatives is about 1.3 ms.

The following note should be made here. It was shown in Para. 2.5 that
construction of the linearized model (2.5.1) includes not only the
matrices of partial derivatives

$$\frac{\partial H}{\partial q}, \quad \frac{\partial C}{\partial q} \quad \text{and} \quad \frac{\partial h_G}{\partial q}$$

COS/SIN	Floating-point multiplications	Floating-point additions/subtractions
4	86	41

Table 5.37. The number of numerical operations needed for computing partial derivatives of model matrices

```
      SUBROUTINE LANTR
      COMMON/LMODL/HL(6,6,6),CL(6,6,6,6),HGL(6,6)
      COMMON/UG/Q(6)
       X8  = Q (2   )
       X2  = C O S ( X8  )
       X5  = S I N ( X8  )
       X9  = Q (3   )
       X3  = C O S ( X9  )
       X6  = S I N ( X9  )
C
      P1=X5*0.16313E+01
      HGL(3,3)=X3*P1
      P1=X6*0.16313E+01
      HGL(3,2)=X2*P1
      P1=X6*0.16313E+01
      HGL(2,3)=X2*P1
      P1=X3*0.163130E+01
      HGL(2,2)=X5*(P1+0.66885E+01)
      P1=X6*0.96400E-01
      Q1=X6*P1
      Q2=X3*(-0.58200E-01)
      CL(2,3,2,3)=Q1+Q2-0.48200E-01
      P1=X3*(-0.85800E-01)
      Q1=X6*(P1-0.58200E-01)
      CL(2,3,1,3)=X5*Q1
      P1=X6*(-0.24000E-02)
      P2=X3*0.40500E-01
      Q1=X6*P1
      Q2=X3*(P2+0.58200E-01)
      CL(2,3,1,2)=X2*(Q1+Q2+0.25500E-02)
      P1=X2*0.58200E-01
      CL(1,3,3,3)=X3*(P1-0.23300E-01)
      P1=X6*0.58200E-01
      CL(1,3,3,2)=X5*P1
      P1=X6*(-0.85800E-01)
      Q1=X5*P1
      CL(1,3,2,3)=X3*Q1
      P1=X3*0.24000E-02
      P2=X6*(-0.40500E-02)
      Q1=X3*P1
      Q2=X6*P2
      CL(1,3,2,2)=X2*(Q1+Q2-0.25500E-02)
      P1=X6*(-0.84000E-01)
      Q1=X3*0.58200E-01
      Q2=X6*P1
      CL(1,2,2,3)=X2*(Q1+Q2+0.43000E-01)
      P1=X3*(-0.43000E-01)
```

Fig. 5.12. Program for forming the matrices of the linearized dynamic model of anthropomorphic robot

218

```
Q1=X6*(P1-0.58200E-01)
CL(1,2,2,2)=X5*Q1
P1=X6*(-0.93000E-02)
P2=X6*0.76500E-01
Q1=X6*P1
Q2=X3*(-0.58200E-01)
Q3=X6*P2
P1=X6*(-0.76500E-01)
P2=X5*(Q1+0.46500E-01)
P3=X3*(-0.23300E-01)
P4=X2*(Q1+Q2-0.38250E-01)
Q1=X6*P1
Q2=X5*P2
Q3=X2*(P3+P4)
CL(1,3,1,3)=Q1+Q2+Q3+0.38250E-01
P1=X3*0.85800E-01
Q1=X2*(P1+0.11640E+00)
P1=X6*(Q1+0.23300E-01)
Q1=X5*P1
CL(1,3,1,2)=Q1
CL(1,2,1,3)=Q1
P1=X3*0.85800E-01
Q1=X5*(P1+0.23280E+00)
P1=X5*0.39410E+00
P2=X3*(-0.42900E-01)
P3=X2*(-0.23300E-01)
P4=X5*Q1
Q1=X5*P1
Q2=X2*(-0.95450E-01)
Q3=X3*(P2+P3+P4)
CL(1,2,1,2)=Q1+Q2+Q3+-0.19705E+00
P1=X3*(-0.85500E-01)
HL(2,2,3)=X6*(P1-0.11640E+00)
P1=X2*(-0.58200E-01)
HL(1,3,3)=X6*(P1-0.23280E-01)
P1=X3*(-0.58200E-01)
HL(1,3,2)=X5*(P1-0.43050E-01)
P1=X6*(-0.85500E-01)
Q1=X6*0.58200E-01
Q2=X6*P1
HL(1,2,3)=X5*(Q1+Q2+0.42750E-01)
P1=X3*0.42750E-01
Q1=X6*(P1+0.58200E-01)
HL(1,2,2)=X2*Q1
P1=X3*(-0.85500E-01)
Q1=X2*(P1+0.11640E+00)
P1=X3*0.85500E-01
P2=X2*(Q1-0.46560E-01)
HL(1,1,3)=X6*(P1+P2)
P1=X3*(-0.85500E-01)
Q1=X3*(P1-0.23280E+00)
P1=X3*(-0.46560E-01)
P2=X2*(Q1-0.39040E+00)
HL(1,1,2)=X5*(P1+P2-0.19090E+00)
RETURN
END
```

Fig. 5.12. (Contd.)

i,j	$E_{H_{ij}}$	$X_k,\ k>6$	$V_{H_{ij}}$
$1,1$	$\begin{bmatrix} 2 & 3 & 2 & 3 & 10 \\ 6 & 2 & 12 & 13 & 0 \\ 7 & 9 & 8 & 0 & 0 \\ 0 & 6 & 11 & 0 & 0 \end{bmatrix}$	$\begin{aligned} X_7 &= 0.22619 \\ X_8 &= 0.19520 \\ X_9 &= 0.19090 \\ X_{10} &= 0.04275 \\ X_{11} &= 0.04275 \\ X_{12} &= 0.04656 \\ X_{13} &= 0.11640 \end{aligned}$	$\begin{bmatrix} 3 & 2 & 2 & 1 & 0 \\ 1 & 0 & 0 & 0 & 0 \\ 0 & 0 & 0 & 0 & 0 \\ 0 & 1 & 0 & 0 & 0 \end{bmatrix}$
$1,2$	$\begin{bmatrix} 5 & 6 & 3 & 7 \\ 0 & 0 & 8 & 0 \end{bmatrix}$	$\begin{aligned} X_7 &= 0.04275 \\ X_8 &= 0.05820 \end{aligned}$	$\begin{bmatrix} 1 & 2 & 1 & 0 \\ 0 & 0 & 0 & 0 \end{bmatrix}$
$1,3$	$\begin{bmatrix} 3 & 2 & 9 \\ 2 & 8 & 0 \\ 0 & 7 & 0 \end{bmatrix}$	$\begin{aligned} X_7 &= 0.04305 \\ X_8 &= 0.02328 \\ X_9 &= 0.05820 \end{aligned}$	$\begin{bmatrix} 2 & 1 & 0 \\ 1 & 0 & 0 \\ 0 & 0 & 0 \end{bmatrix}$
$2,2$	$\begin{bmatrix} 3 & 3 & 8 \\ 7 & 9 & 0 \end{bmatrix}$	$\begin{aligned} X_7 &= 0.19920 \\ X_8 &= 0.04275 \\ X_9 &= 0.11640 \end{aligned}$	$\begin{bmatrix} 2 & 1 & 0 \\ 2 & 1 & 0 \end{bmatrix}$

i	$k,$	$E_{C^i_{k\ell}}$	$X_k,\ k>6$	$V_{C^i_{k\ell}}$
1	$2,1$	$\begin{bmatrix} 5 & 3 & 2 & 3 & 11 \\ 0 & 2 & 10 & 9 & 0 \\ 0 & 7 & 8 & 0 & 0 \end{bmatrix}$	$\begin{aligned} X_7 &= -0.09545 \\ X_8 &= -0.19705 \\ X_9 &= -0.11640 \\ X_{10} &= -0.02330 \\ X_{11} &= -0.04290 \end{aligned}$	$\begin{bmatrix} 3 & 2 & 2 & 1 & 0 \\ 0 & 1 & 0 & 0 & 0 \\ 0 & 0 & 0 & 0 & 0 \end{bmatrix}$

Table 5.38. Matrix representatives of the elementary graphs
of the dynamic model matrices

		$E_{h_i}^G$		$V_{h_i}^G$
1	3,1	$\begin{bmatrix} 6 & 2 & 2 & 3 & 10 \\ 0 & 3 & 8 & 9 & 11 \\ 0 & 0 & 5 & 5 & 0 \end{bmatrix}$	$\begin{aligned} X_7 &= 0.03825 \\ X_8 &= -0.02330 \\ X_9 &= -0.05820 \\ X_{10} &= -0.03825 \\ X_{11} &= 0.00465 \end{aligned}$	$\begin{bmatrix} 2 & 2 & 2 & 1 & 0 \\ 0 & 1 & 0 & 0 & 0 \\ 0 & 0 & 1 & 1 & 0 \end{bmatrix}$
1	2,2	$\begin{bmatrix} 2 & 6 & 3 & 8 \\ 0 & 0 & 7 & 0 \end{bmatrix}$	$\begin{aligned} X_7 &= 0.05820 \\ X_8 &= 0.04300 \end{aligned}$	$\begin{bmatrix} 1 & 2 & 1 & 0 \\ 0 & 0 & 0 & 0 \end{bmatrix}$
1	3,2	$\begin{bmatrix} 5 & 6 & 6 & 7 \\ 0 & 3 & 3 & 9 \\ 0 & 8 & 0 & 0 \end{bmatrix}$	$\begin{aligned} X_7 &= -0.04050 \\ X_8 &= -0.00255 \\ X_9 &= 0.00240 \end{aligned}$	$\begin{bmatrix} 3 & 1 & 1 & 0 \\ 0 & 1 & 1 & 0 \\ 0 & 0 & 0 & 0 \end{bmatrix}$
1	3,3	$\begin{bmatrix} 6 & 2 & 8 \\ 0 & 7 & 0 \end{bmatrix}$	$\begin{aligned} X_7 &= -0.02330 \\ X_8 &= -0.05820 \end{aligned}$	$\begin{bmatrix} 2 & 1 & 0 \\ 0 & 0 & 0 \end{bmatrix}$
2	3,1	$\begin{bmatrix} 5 & 3 & 3 & 8 \\ 0 & 6 & 7 & 10 \\ 0 & 9 & 6 & 0 \end{bmatrix}$	$\begin{aligned} X_7 &= 0.05820 \\ X_8 &= 0.04050 \\ X_9 &= 0.00255 \\ X_{10} &= -0.00240 \end{aligned}$	$\begin{bmatrix} 3 & 2 & 1 & 0 \\ 0 & 1 & 0 & 0 \\ 0 & 0 & 1 & 0 \end{bmatrix}$
2	3,2	$\begin{bmatrix} 6 & 3 & 8 \\ 0 & 7 & 0 \end{bmatrix}$	$\begin{aligned} X_7 &= -0.0582 \\ X_8 &= -0.0428 \end{aligned}$	$\begin{bmatrix} 2 & 1 & 0 \\ 0 & 0 & 0 \end{bmatrix}$
i		$E_{h_i}^G$		$V_{h_i}^G$
2		$\begin{bmatrix} 2 & 3 & 8 \\ 0 & 7 & 0 \end{bmatrix}$	$\begin{aligned} X_7 &= -6.68850 \\ X_8 &= -1.63130 \end{aligned}$	$\begin{bmatrix} 2 & 1 & 0 \\ 0 & 0 & 0 \end{bmatrix}$
3		$\begin{bmatrix} 5 & 6 & 7 \end{bmatrix}$	$X_7 = 1.63130$	$\begin{bmatrix} 1 & 1 & 0 \end{bmatrix}$

Table 5.38. (Contd.)

but also the matrices of nonlinear model $H(q, \theta)$ and $C(q, \theta)$. This is
why it is necessary to determine the elementary graphs of the terms of
matrices H and C. Matrix representatives of the graphs are given in
Table 5.38. The number of multiplications and additions required to
form matrices $H(q, \theta)$ and $C(q, \theta)$ may easily be determined from these
results. Thus, one obtains the data given in Table 5.39. As may be seen
from this table, the number of operations for computing the matrices
of nonlinear model is about

$$n_{nonlin} \cong 50\% \; n_{lin},$$

where n_{lin} is the number of operations for forming the partial derivatives of these matrices. Thus, it follows that the number of operations
required for the complete linearized model is about three times larger
than that for forming the nonlinear model.

5.5. Microcomputer Implementation of Analytical Robot Models – Time-Memory Requirements

In this section we will consider implementation of analytical robot
models on up-to-date microcomputers from the point of view of the required memory resources and operation speeds, not only for three - but
also for six-degree-of-freedom manipulators. To ensure the results to
have their applicative value, measurements have been made on typical
industrial robots. Robots of cylindrical, arthropoid and anthropomorphic structure with three joints have been considered in the preceding
text. The analysis will now be extended to include also typical six-
degree-of-freedom robots which are commonly used in practice. For example, we will consider a robot whose first three degrees of freedom are
of cylindrical type, and the remaining three of anthropomorphic structure (Fig. 5.13). In addition, we will consider a well-known robot
("Stanford manipulator") which is frequently referred to as an example
in dynamic analysis and robot control synthesis [33], Fig. 5.14. Finally, we will observe a robot whose structure is more complex than the
all the preceding ones (Semiarthropoidal robot), Fig. 5.15. Such robots
are suitable for welding and painting tasks.

The EXPERT-PROGRAM, whose structure was described in detail in Ch. 4,
was first used to generate the program codes for analytical models of
the robots mentioned. As previously stated, the expert-program as well

222

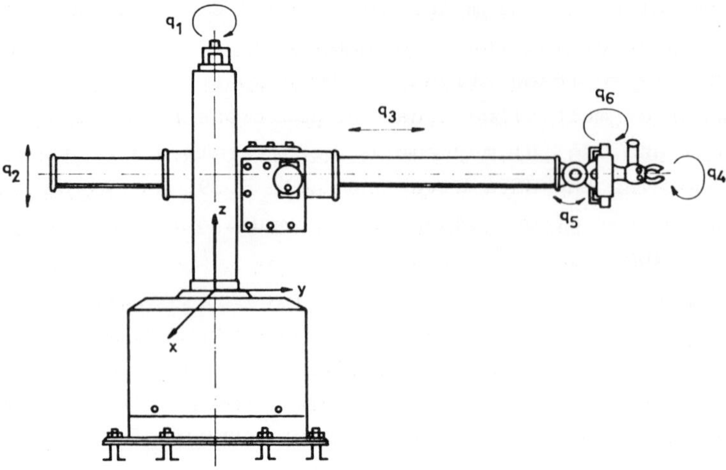

Fig. 5.13. Cylindrical robot with 6 degrees of freedom (CL-AN)

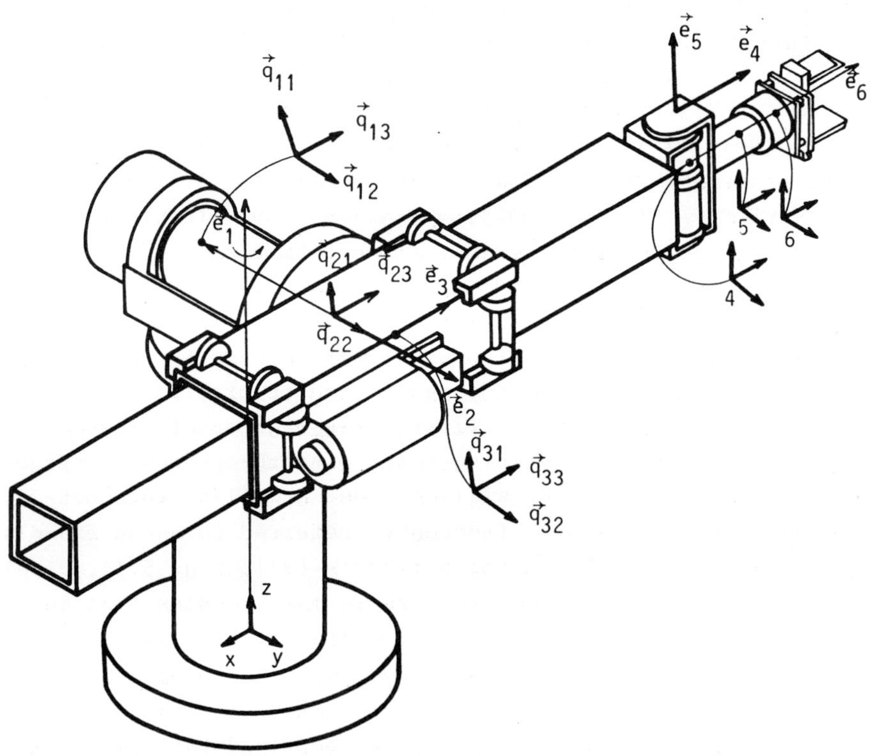

Fig. 5.14. "Stanford manipulator" with 6 degrees of freedom (ST)

as other of the developed program system for robot modelling are suitable for installation on high-performance computers. The obtained programs for real-time operation should then be installed on appropriate
microcomputers. In this actual case, model generation was performed on
PDP 11/70 computer, and testing of the obtained models on microcomputers INTEL 8086 (with 8087 arithmetic coprocessor), 16-bit microcomputer
PDP 11/03 and on PDP 11/70 itself. Compilation of source programs and
their testing were performed in microcomputers. Appendix 5.1 presents
an example of a generated source program. This program represents the
model of Stanford manipulator with 6 degrees of freedom. As may be seen,
the number of floating-point multiplications and additions/subtractions
is about 7.5 times larger than the corresponding number of operations
required for the most complex manipulator with three degrees of freedom
(anthropomorphic).

For each of the mentioned manipulators, a number of important parameters relating to the possible real-time implementation of their dynamic
models were measured. Number of floating-point operations and the time
of calculating the model matrices are systematized in Table 5.39. The
first column of the table contains shortened names of robots (CL - cylindrical robot, AR - arthropoid, AN - anthropomorphic, sAR-AN semiarthropoidal robot, ST - Stanford manipulator), and the second column the
corresponding kinematic schemes and joint coordinates of robots. As may
be seen, less than 60 floating-point operations are required for manipulators with three degrees of freedom, and several hundred operations
for manipulators with six degrees of freedom. A graphical representation of the numbers of operations as function of the complexity of the
kinematic scheme is provided in Fig. 5.16. The shortened robot names
from Table 5.39 are given on the abscissa. This table also contains
the computer times for calculating all the terms of model matrices,
obtained by measurements on micro/minicomputers. We can now draw, a
number of very important conclusions about the possibilities for real-
-time implementation of dynamic robot models. As may be seen, all models can be formed in real time on PDP 11/70 minicomputer. If cosine
and sine functions are calculated using table look-up method (rows
marked by "T" in Table 5.39), it may be seen that less than 0.6 ms is
required to form the dynamic models of even the most complex kinematic
schemes of robots with three degrees of freedom. If trigonometric functions are calculated using library programs, less than 1.2 ms is required. It should be stated that it is usually not necessary to use
library programs (with calculation accuracy of 6 and more significant
digits), since the kinematic and dynamic parameters of robots are

Table 5.39

		cos/ /sin	number of multipl.	number of additions	L/T*)	t[ms]		
						PDP11/70	INTEL 86	PDP11/03
CL	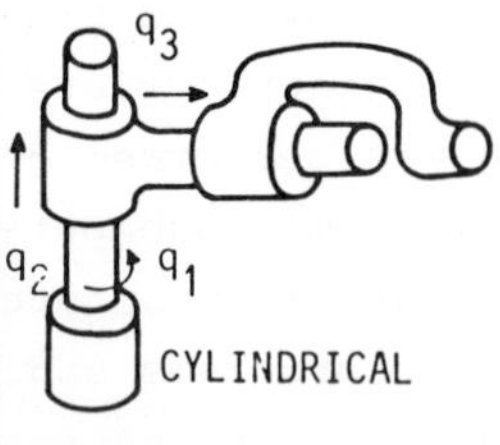 q3 q2 q1 CYLINDRICAL	0	3	3	T	0.2	0.9	1.7
					L	0.2	0.9	1.7
AR	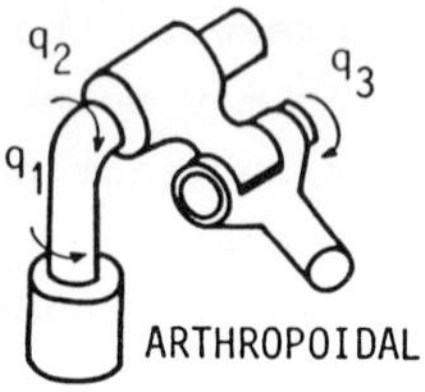 q2 q3 q1 ARTHROPOIDAL	4	49	22	T	0.5	4.4	11.4
					L	0.97	6.4	52.0
AN	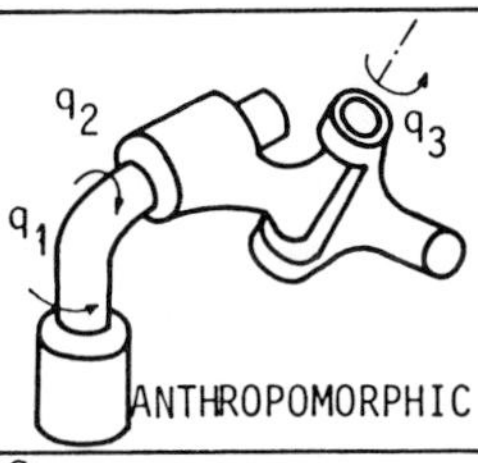 q2 q3 q1 ANTHROPOMORPHIC	4	55	31	T	0.6	5.3	13.6
					L	1.2	7.4	54.0
CL-AN	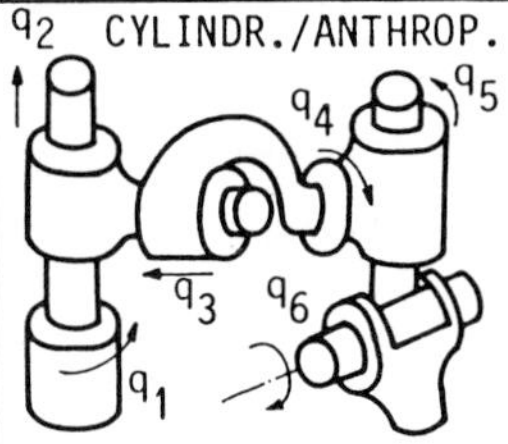 q2 CYLINDR./ANTHROP. q4 q5 q3 q6 q1	6	193	80	T	2.8	16.9	41.5
					L	3.2	19.9	107.5
ST	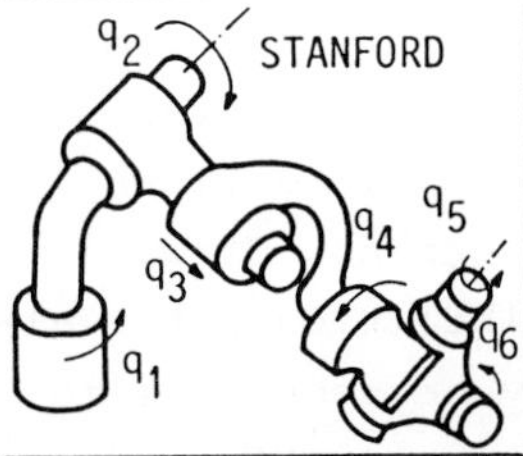 q2 STANFORD q4 q5 q3 q6 q1	6	372	167	T	5.0	30.2	76.5
					L	5.4	33.3	145.0
sAR-AN	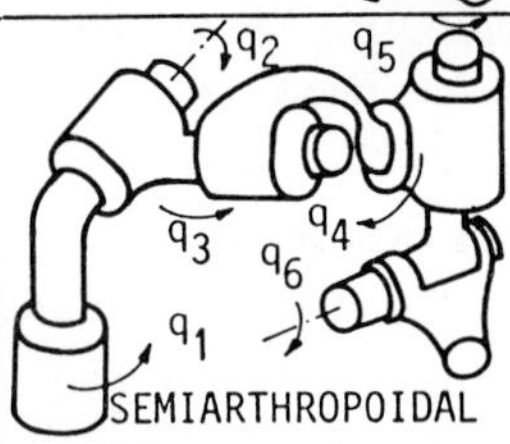 q2 q5 q3 q4 q6 q1 SEMIARTHROPOIDAL	8	895	389	T	13.9	83.7	209.0
					L	14.4	88.0	297.0

*) T means that trigonometric functions are calculated by using table-look-up method and linear interpolation, L - using the library.

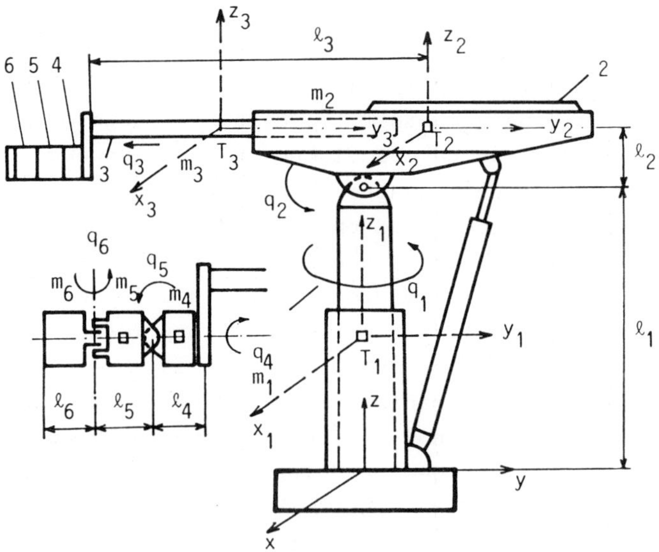

Fig. 5.15. Semiarthropoidal robot with 6 degrees of freedom
(sAR-AN)

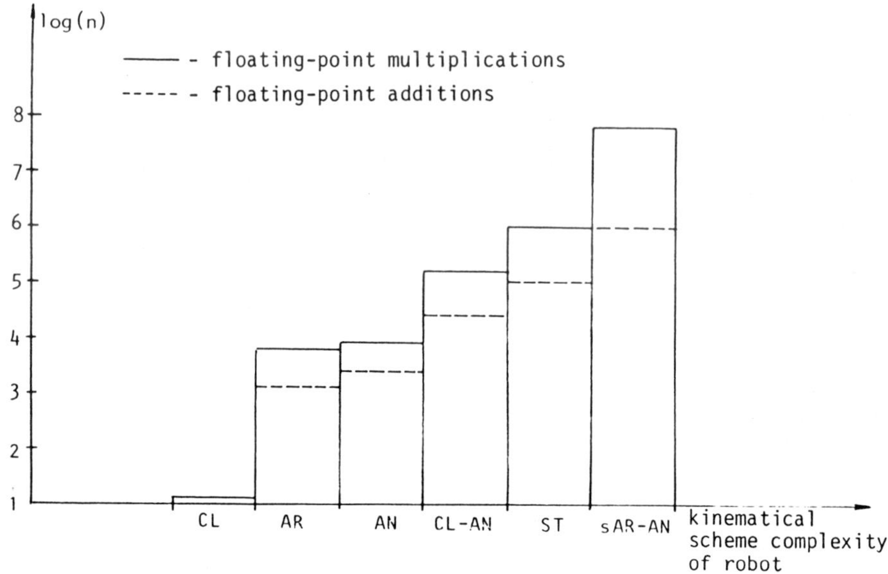

Fig. 5.16. Number of floating-point operations as a function of
the complexity of manipulator kinematic scheme (log –
a natural logarithm, n – the number of operations
from Table 5.39)

oftenly known with only 2 to 3 significant digits. The time to form the model of six-degree-of-freedom manipulator ranges from 3 to 14 ms. If 20-30 ms is accepted as real-time limit, which corresponds to 30-50 Hz sampling frequency, it may be seen that the dynamic models of robots with six degrees of freedom can be calculated on-line. However, the times achievable by microcomputer systems are much more important for practical applications. A system based on INTEL 8086 microprocessor (with a coprocessor) and PDP 11/03 is considered in Table 5.39. It follows from the table that the first microcomputer is 2-3 times faster than the second, so the possibility for real-time application of dynamic robot models will be analyzed on INTEL 8086. It may be seen from the table that the dynamic models of manipulators with three degrees of freedom can be formed in less than 5 ms. Let us note that the application of numerical methods (e.g., Newton-Euler's method) under identical conditions requires more than 20 ms. The time of calculating the terms of model matrices of 5 ms is very attractive, since it accounts for only 25% of the time available at 50 Hz sampling frequency. The remaining 75% of the time can be used to form manipulator trajectories in the space of joint coordinates, to process sensor signals, etc. As far as six-degree-of-freedom manipulators are concerned, the models of cylindrical-anthropomorphic robot (about 20 ms), and of the Stanford manipulator (about 30 ms) can be formed in real time. The time of about 80 ms is required to form the model of the most complex robot (sAR-AN in Table 5.39). To enable real-time implementation of such a model, certain simplifications should be introduced into the model (see Para. 3.7). In addition, it is possible to use parallel processing or faster microprocessors (e.g., INTEL 286 and so on).

Fig. 5.17 provides a graphical representation of the times to calculate the matrices of dynamic robot models as functions of the complexity of their kinematic schemes. The real-time limit of 20-30 ms is also shown.

Apart from program execution speeds, we will also consider memory requirements for implementation of the dynamic models of robots. Since these data are difficult to foresee, the sizes presented here are those obtained by measurements performed on actual computer systems. Measurements included the sizes of memory regions required for storing the source programs and for storing the object programs in machine code. These data are expressed in units of 8 bits each (1 byte) and are shown in Table 5.40 for all mentioned robots. As may be seen, a typical microcomputer suitable for robot control (e.g., INTEL 8086) requires about 2000 bytes for the source programs of the models of robots with

3 joints and about 8000 to 30000 bytes for robots with six degrees of freedom. The memory space required for installing the object programs of robot models is up to 5 Kbytes for manipulators with no more than 3 degrees of freedom, and from 13 to 58 Kbytes for manipulators with six degrees of freedom. The size of the required memory space as a function of the complexity of the kinematic scheme of manipulators is shown in Fig. 5.18. Finally, it should be noted that, in view of the comparatively low price of today's semiconductor memories, this issue is considerably less significant than that of microcomputer operation speed, i.e. the possibility for real-time implementation of robot models.

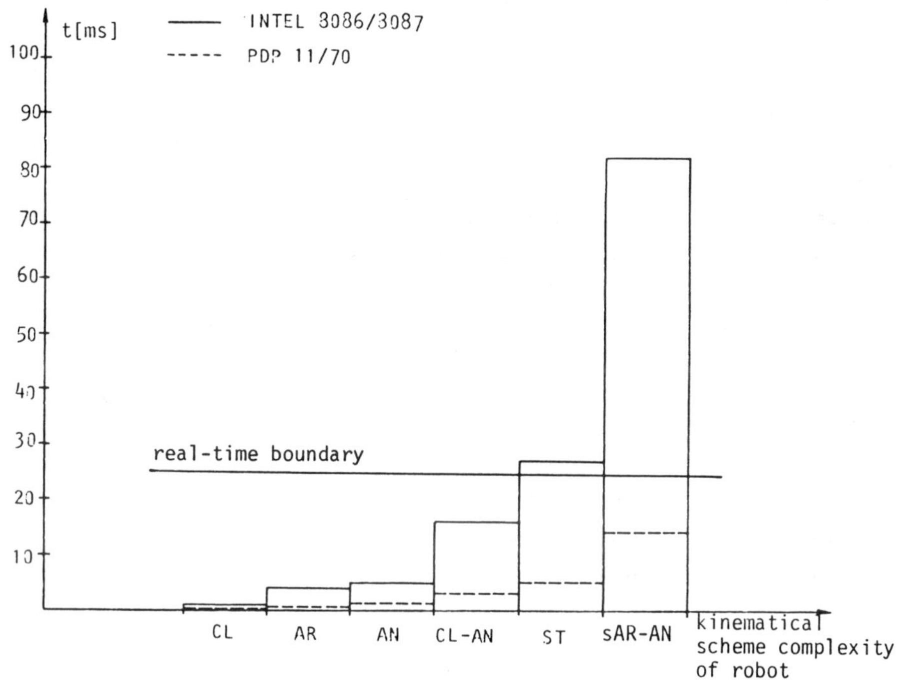

Fig. 5.17. Calculation time of robot dynamic models matrices as a function of the complexity of their kinematic schemes

ROBOT PROGRAM	Memory (byte), INTEL 8086/8087					
	CL	AR	AN	CL-AN	ST	sAR-AN
Source program	1948	2184	2385	8041	11644	30493
Computer code	1219	3809	5190	13509	21925	58210

Table 5.40. Memory required to install the program for calculating the matrices of dynamic robot models

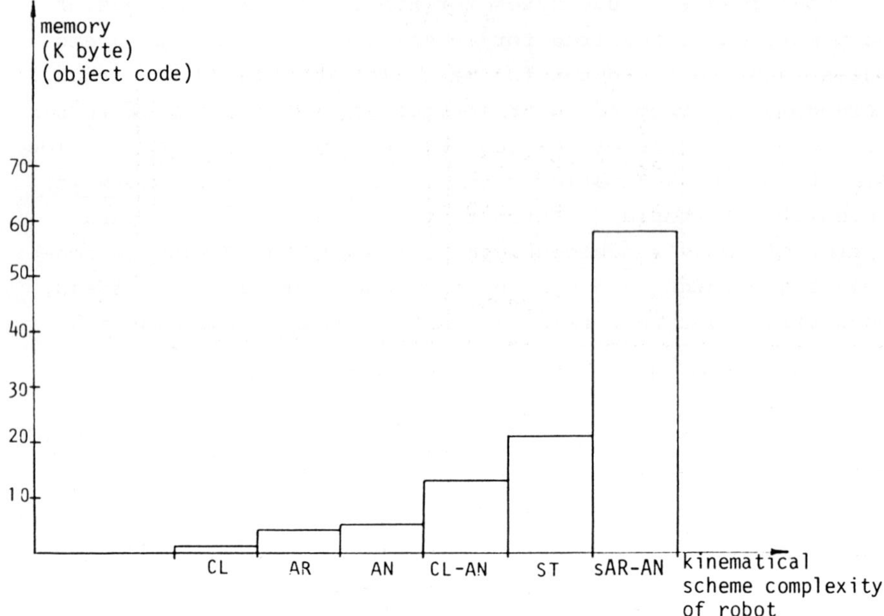

Fig. 5.18. Memory requirements of the programs for dynamic
models computation versus robot kinematic complexity

Appendix 5.1

In this appendix we will present the program of the dynamic model of
Stanford-manipulator with six degrees of freedom whose kinematic scheme
is shown in Table 5.39. This program was obtained as the output file of
expert program.

```fortran
      SUBROUTINE STANFM
      COMMON/MATRIC/H(6,6),C(6,6,6),G(6)
      COMMON/UG/Q(6)
       X11 = Q (1   )
       X12 = Q (2   )
       X2  = C O S ( X12 )
       X7  = S I N ( X12 )
       X13 = Q (3   )
       X14 = Q (4   )
       X4  = C O S ( X14 )
       X9  = S I N ( X14 )
       X15 = Q (5   )
       X5  = C O S ( X15 )
       X10 = S I N ( X15 )
C
      Q1=X4*(-0.13773E+01)
      P2=X5*Q1
      P1=X2*(-0.13773E+01)
      Q2=X7*P2
      Q1=X10*P1
      G(5)=Q1+Q2
      Q1=X7* 0.13773E+01
      P1=X9*Q1
      Q1=X10*P1
      G(4)=Q1
      Q1=X2* 0.63471E+02
      G(3)=Q1
      Q1=X2*(-0.13773E+01)
      P3=X4*Q1
      P2=X5*(-0.13773E+01)
      P1=X13*(-0.63471E+02)
      Q2=X10*P3
      Q1=X7*(P1+P2 -0.14047E+02)
      G(2)=Q1+Q2
      G(1)=0.
      P1=X5*(-0.27697E-01)
      P2=X5* 0.16000E 01
      P1=X2* 0.12397E-01
      Q4=X5*P2
      Q3=X4*P1
      Q2=X5* 0.27697E-01
      Q1=X13* 0.14040E+00
```

```
P4=X4*(Q4 -0.15000E-01)
P3=X4*(-0.12397E-01)
P2=X10*Q3
P1=X7*(Q1+Q2+ 0.90556E-01)
Q3=X7*P4
Q2=X10*P3
Q1=X2*(P1+P2)
P3=X4*(-0.21397E-01)
P2=X7*Q3
P1=X10*(Q1+Q2)
Q2=X10*P3
Q1=X9*(P1+P2)
C(4,1,1)=Q1+Q2
Q1=X5*(-0.14040E+00)
C(3,5,5)=Q1
P1=X5* 0.15300E-01
Q1=X10*P1
C(3,5,4)=Q1
C(3,4,4)=0.
P1=X5*(-0.16000E-01)
Q1=X5*P1
P1=X4*(Q1+ 0.15000E-01)
Q3=X5*(-0.14040E+00)
Q2=X13*(-0.64700E+01)
Q1=X9*P1
C(3,2,2)=Q1+Q2+Q3 -0.14319E+01
Q2=X5* 0.80000E-02
Q1=X5*(-0.80000E-02)
P3=X2* 0.15300E-01
P2=X5*Q2
P1=X5*Q1
Q4=X4*P3
Q3=X10*(-0.80000E-02)
Q2=X4*(P2 -0.75000E-02)
Q1=X9*(P1+ 0.75000E-02)
P4=X5*Q4
P3=X10*Q3
P2=X4*Q2
P1=X9*Q1
Q2=X10*P4
Q1=X7*(P1+P2+P3 -0.35000E-02)
C(3,2,1)=Q1+Q2
P1=X5* 0.16000E-01
Q2=X5* 0.15300E-01
Q1=X5*P1
P3=X4*(-0.14040E+00)
P2=X9*Q2
P1=X4*(Q1 -0.15000E-01)
Q4=X2*(P2+P3)
Q3=X5*(-0.14040E+00)
Q2=X13*(-0.64700E+01)
Q1=X9*P1
P2=X10*Q4
P1=X7*(Q1+Q2+Q3 -0.14319E+01)
Q1=X7*(P1+P2)
C(3,1,1)=Q1
Q1=X13*(-0.14040E+00)
P1=X4*(Q1 -0.90556E-01)
Q1=X10*P1
C(2,5,5)=Q1

Q3=X10* 0.76500E-02
Q2=X5*(-0.20047E-01)
Q1=X13*(-0.14040E+00)
P2=X10*Q3
P1=X5*(Q1+Q2 -0.90556E-01)
Q1=X9*(P1+P2 -0.80000E-02)
C(2,5,4)=Q1
C(2,5,3)=0.
Q2=X5*(-0.27697E-01)
Q1=X13*(-0.14040E+00)
P1=X4*(Q1+Q2 -0.90556E-01)
Q1=X10*P1
C(2,4,4)=Q1
C(2,4,3)=0.
C(2,3,3)=0.
P2=X4* 0.11697E-01
P1=X4*(-0.16000E-01)
Q7=X13*(-0.14040E+00)
Q6=X5*(-0.27697E-01)
Q5=X13*(-0.14040E+00)
Q4=X5* 0.27697E-01
Q3=X13* 0.14040E+00
Q2=X4*P2
Q1=X4*P1
P8=X2*(X5*(-0.27700E-01)+Q7)
P7=X2*(Q5+Q6 -0.90556E-01)
P6=X4*(Q3+Q4+ 0.90556E-01)
P5=X9*(-0.17000E-01)
P4=X10*(Q2+ 0.16000E-01)
P3=X13*(-0.64700E+01)
P2=X13*(-0.28079E+00)
P1=X5*(Q1 -0.11697E-01)
Q7=X2*P8
Q6=X7*P7
Q5=X4*P6
Q4=X9*P5
Q3=X10*P4
Q2=X13*(P3 -0.28638E+01)
Q1=X5*(P1+P2 -0.18111E+00)
P3=X4*Q7
P2=X10*Q6
P1=X2*(Q1+Q2+Q3+Q4+Q5 -0.33266E+01)
Q2=X10*P3
Q1=X7*(P1+P2)
C(2,1,1)=Q1+Q2
P1=X13*(-0.14040E+00)
Q2=X2* 0.21397E-01
Q1=X7*(P1 -0.90556E-01)
P3=X5* 0.21397E-01
P2=X4*Q2
P1=X9*Q1
Q2=X7*P3
Q1=X10*(P1+P2)
C(1,5,5)=Q1+Q2
P3=X10*(-0.76500E-02)
P2=X5* 0.20047E-01
P1=X13* 0.14040E+00
Q4=X9* 0.21397E-01
Q3=X10* 0.27697E-01
Q2=X10*P3
Q1=X5*(P1+P2+ 0.90556E-01)
```

```
P2=X2*(Q3+Q4)
P1=X4*(Q1+Q2+ 0.80000E-02)
Q2=X5*P2
Q1=X7*P1
C(1,5,4)=Q1+Q2
C(1,5,3)=0.
Q1=X4* 0.27697E-01
P3=X5*Q1
P2=X5*(-0.76500E-02)
P1=X10* 0.20047E-01
Q4=X4* 0.21397E-01
Q3=X7*P3
Q2=X5*P2
Q1=X10*P1
P4=X5*Q4
P3=X2* 0.21397E-01
P2=X10*Q3
P1=X2*(Q1+Q2+ 0.80000E-02)
Q3=X7*P4
Q2=X10*P3
Q1=X9*(P1+P2)
C(1,5,2)=Q1+Q2+Q3
P2=X5*(-0.27697E-01)
P1=X13*(-0.14040E+00)
Q2=X2* 0.21397E-01
Q1=X7*(P1+P2 -0.90556E-01)
P2=X4*Q2
P1=X9*Q1
Q1=X10*(P1+P2)
C(1,4,4)=Q1
C(1,4,3)=0.
Q3=X5*(-0.80000E-02)
Q2=X5* 0.80000E-02
Q1=X10*(-0.12047E-01)
P3=X5*Q3
P2=X5*Q2
P1=X10*Q1
Q4=X10*(-0.80000E-02)
Q3=X4*(P3+ 0.75000E-02)
Q2=X10*(-0.21397E-01)
Q1=X9*(P1+P2 -0.75000E-02)
P3=X10*Q4
P2=X4*Q3
P1=X9*(Q1+Q2)
Q1=X7*(P1+P2+P3)
C(1,4,2)=Q1
C(1,3,3)=0.
Q1=X2*(-0.98603E+00)
C(1,3,2)=Q1
Q2=X5*(-0.16000E-01)
Q1=X2* 0.11697E-01
P4=X5*Q2
P3=X4*Q1
P2=X5* 0.27697E-01
P1=X13* 0.14040E+00
Q4=X2* 0.21397E-01
Q3=X2*(P4+ 0.15000E-01)
Q2=X10*P3
Q1=X7*(P1+P2+ 0.90556E-01)
P5=X4*Q4
P4=X5* 0.21397E-01
P3=X13* 0.98603E+00

P2=X4*Q3
P1=X10*(Q1+Q2)
Q3=X10*P5
Q2=X7*(P3+P4+ 0.21822E+00)
Q1=X9*(P1+P2)
C(1,2,2)=Q1+Q2+Q3
H(5,5)=    0.28397E-01
H(5,4)=0.
Q1=X10*(-0.14040E+00)
H(5,3)=Q1
Q1=X13* 0.14040E+00
P1=X5*(Q1+ 0.90556E-01)
Q1=X4*(P1+ 0.28397E-01)
H(5,2)=Q1
P1=X13* 0.14040E+00
Q2=X2*(-0.21397E-01)
Q1=X5*(P1+ 0.90556E-01)
P3=X4*Q2
P2=X10* 0.21397E-01
P1=X7*(Q1+ 0.28397E-01)
Q2=X5*P3
Q1=X7*(P1+P2)
H(5,1)=Q1+Q2
P1=X10* 0.28397E-01
Q1=X10*P1
H(4,4)=Q1+ 0.10000E-02
H(4,3)=0.
Q2=X5*(-0.27697E-01)
Q1=X13*(-0.14040E+00)
P1=X9*(Q1+Q2 -0.90556E-01)
Q1=X10*P1
H(4,2)=Q1
P2=X5* 0.27697E-01
P1=X13* 0.14040E+00
Q3=X9* 0.21397E-01
Q2=X10* 0.28397E-01
Q1=X4*(P1+P2+ 0.90556E-01)
P2=X2*(Q2+Q3)
P1=X7*Q1
Q2=X2* 0.10000E-02
Q1=X10*(P1+P2)
H(4,1)=Q1+Q2
H(3,3)=    0.64700E+01
P1=X4*(-0.14040E+00)
Q1=X10*P1
H(3,2)=Q1
Q1=X9*(-0.14040E+00)
P1=X10*Q1
Q1=X7*(P1 -0.98603E+00)
H(3,1)=Q1
P2=X10* 0.12397E-01
P1=X9* 0.16000E-01
Q2=X10*P2
Q1=X9*P1
P5=X9* 0.20000E-02
P4=X4*(Q2+ 0.17000E-01)
P3=X13* 0.64700E+01
P2=X13* 0.28079E+00
P1=X5*(Q1+ 0.12397E-01)
Q4=X9*P5
Q3=X4*P4
Q2=X13*(P3+ 0.28638E+01)
```

```
Q1=X5*(P1+P2+ 0.18111E+00)
H(2,2)=Q1+Q2+Q3+Q4+ 0.34516E+01
Q2=X5*(-0.16000E-01)
Q1=X4* 0.11697E-01
P4=X5*Q2
P3=X7*Q1
P2=X5*(-0.27697E-01)
P1=X13*(-0.14040E+00)
Q4=X4* 0.21397E-01
Q3=X4*(P4+ 0.15000E-01)
Q2=X10*P3
Q1=X2*(P1+P2 -0.90556E-01)
P5=X7*Q4
P4=X5*(-0.21397E-01)
P3=X13*(-0.98603E+00)
P2=X7*Q3
P1=X10*(Q1+Q2)
Q3=X10*P5
Q2=X2*(P3+P4 -0.21822E+00)
Q1=X9*(P1+P2)
H(2,1)=Q1+Q2+Q3
P2=X4* 0.12397E-01
P1=X4* 0.16000E-01
Q4=X4*P2
Q3=X5* 0.55394E-01
Q2=X13* 0.28079E+00
Q1=X4*P1
P8=X9* 0.12397E-01
P7=X2*(Q4+ 0.16000E-01)
P6=X2*(Q2+Q3+ 0.18111E+00)
P5=X4* 0.20000E-02
P4=X9* 0.17000E-01
P3=X13* 0.64700E+01
P2=X13* 0.28079E+00
P1=X5*(Q1+ 0.12397E-01)
Q7=X9*P8
Q6=X2*P7
Q5=X4*P6
Q4=X4*P5
Q3=X9*P4
Q2=X13*(P3+ 0.28638E+01)
Q1=X5*(P1+P2+ 0.18111E+00)
P5=X2* 0.10700E+00
P4=X9* 0.42793E-01
P3=X10*(Q6+Q7)
P2=X10*Q5
P1=X7*(Q1+Q2+Q3+Q4+ 0.34336E+01)
Q3=X2*P5
Q2=X10*(P3+P4)
Q1=X7*(P1+P2)
H(1,1)=Q1+Q2+Q3+ 0.41634E+00

RETURN
END
```

References

[1]   Uicker J.J., On Dynamic Analysis of Spatial Linkages Using 4×4 Matrices, Ph. D. dissertation, Northwestern University, Evanstone, August, 1965.

[2]   Uicker J.J., "Dynamic Force Analysis of Spatial Linkages", Lafayette, Ind., Oct. 10-12, 1966.

[3]   Uicker J.J., "Dynamic Force Analysis of Spatial Linkages", ASME Journal of Applied Mechanics, June, pp. 418-424, 1967.

[4]   Kahn M.E., The Near Minimum Time Control of Open Loop Articulated Kinematic Chains, Ph. D. Thesis, Stanford University, MEMO AIM. 106, 1969.

[5]   Kahn M.E., Roth B., "The Near-Minimum-Time Control of Open-loop Articulated Kinematic Chains" ASME Journal of Dynamic Systems, Measurement and Control, Vol. 93, pp. 164-172, 1971.

[6]   Ball R.S., A Treatise on the Theory of Screws, Cambridge University press, London, 1900.

[7]   Brand L., Vector and Tensor Analysis, 4th ed., New York, Wiley.

[8]   Woo L.S., Freudenstein F., "Dynamic Analysis of Mechanisms Using Screw Coordinates", ASME Journal of Engineering for Industry, February, 1971.

[9]   Yang A.T., "Inertia Force Analysis of Spatial Mechanisms", ASME Journal of Engineering for Industry, February, 1971.

[10] Orlandea N., Berenui T., "Dynamic Continuous Path Synthesis of Industrial Robots Using Adams Computer Program", ASME Journal of Mechanical Design, No. 5, 1981.

[11] Mahil S.S., Modeling Techniques in the Control of Anthropomorphic Industrial Manipulators, Ph. D. thesis, Lanchester Polytechnic, Coventry, 1980.

[12] Mahil S.S., "On the Application of Lagrange's Method to the Description of Dynamic Systems", IEEE Trans. on SMC, Vol. 12, No. 6, pp. 877-890, 1982.

[13] Renaud M., "Contribution a 1-Etude de la Modelisation et de la Commande des Systèmes Mécaniques Articulés", Thèse de Docteur-Ingénieur, Toulouse, 1975.

[14] Thomas M., Tesar D., "Dynamic Modeling of Serial Manipulator Arms", ASME Journal of Dynamic Systems, Measurement and Control, Vol. 104, No. 3, pp. 218-228, 1982.

[15] Waters R.C., Mechanical Arm Control, A.I. Memo 549, MIT Artificial Intelligence Laboratory, 1979.

[16] Hollerbach J.M., An Iterative Lagrangian Formulation of Manipulator Dynamics, A.I. Memo 533, MIT Artificial Intelligence Laboratory, March 1980.

234

[17] Hollerbach J.M., "A Recursive Formulation of Lagrangian Manipu-
      lator Dynamics", IEEE Trans. on SMC, Vol. 10, No. 11, pp. 730-
      -736, 1980.

[18] Denavit J., Hartenberg R.S., "A Kinematic Notation for Lower Pair
      Mechanisms Based on Matrices", ASME Journal of Applied Mechanics,
      June, 1955.

[19] Megahed S., Renaud M., "Minimization of the Computation Time Nec-
      essary for the Dynamic Control of Robot Manipulators", 12th ISIR,
      Paris, 1982.

[20] Vukobratović M., Stepanenko Y., "Mathematical Models of General
      Anthropomorphic Systems", Mathematical Biosciences, Vol. 17, pp.
      191-242, 1973.

[21] Stepanenko Y., Vukobratović M., "Dynamics of Articulated Open
      Chain Active Mechanisms", Mathematical Biosciences, Vol. 28, pp.
      137-170, 1976.

[22] Vukobratović M., "Computer Method for Dynamic Model Construction
      of Active Articulated Mechanisms Using Kinetostatic Approach",
      Journal of Mechanism and Machine Theory, Vol. 13, pp. 19-39, 1978.

[23] Vukobratović M., Kirćanski M., "Correlation Between Denavit-Har-
      tenberg Manipulator Kinematics and Rodrigues Formula Approach,
      Journal of Mechanisms and Machine Theory, (in press) 1984.

[24] Renaud N., "An Efficient Iterative Analytical Procedure for Ob-
      taining a Robot Manipulator Dynamic Model", Proc. of First Inter-
      national Symp. of Robotics Research, Bretton Woods, New Hampshire,
      USA, 1983.

[25] Vukobratović M., Potkonjak V., "Contribution to Automatic Forming
      of Active Chain Models via Lagrangian Form", ASME Journal of Ap-
      plied Mechanics, No. 1, 1979.

[26] Vukobratović M., Potkonjak V., Dynamics of Manipulation Robots,
      Springer-Verlag, 1982.

[27] Orin D., McGhee R., Vukobratović M., Hartoch G., "Kinematic and
      Kinetic Analysis of Open-Chain Linkages Utilizing Newton-Euler
      Methods", Mathematical Biosciences, pp. 107-130, 1979.

[28] Cvetković V., Contribution to the Dynamic Control Synthesis of
      Manipulators, Ph. D. Thesis, Electrical Engineering Faculty, Bel-
      grade, 1981.

[29] Cvetković V., Vukobratović M., "Computer-Oriented Algorithm Mod-
      elling of Active Spatial Mechanisms for Application in Robotics",
      IEEE Trans. on SMC, Vol. SMC-12, No. 6, 1982.

[30] Huston R.L., Kelly F.A., "The Development of Equations of Motion
      of Single Arm Robots", IEEE Trans. on SMC, Vol. 12, No. 3, pp.
      259-266, 1982.

[31] Kane T.R., Wang C.F., "On the Derivation of Equations of Motion",
      J. Soc. for Ind. and Appl. Math., Vol. 13, pp. 487-492, 1965.

[32] Kane T.R., Dynamics, New York: Holt, Reinhart, and Winston, 1968.

[33] Luh J.Y.S., Walker M.W., Paul R.P.C., "On-Line Computational Scheme for Mechanical Manipulators", ASME Journal of Dynamic Systems, Measurement and Control, Vol. 102, No. 2, pp. 69-76, 1980.

[34] Medvedov V.S., Leskov V.S., Yuschenko A.C., Control Systems of Manipulation Robots, "Nauka", Moscow, 1978.

[35] Popov E.P., Vereschagin A.F., Zenkevitch S.L., Manipulation Robots: Dynamics and Algorithms, "Nauka", Moscow, 1978.

[36] Kirćanski N., Vukobratović M., "Control of Manipulators by Means of Discrete Regulators", Trans. of the Institute of Measurement and Control, Vol. 2, No. 3, 1980.

[37] Vukobratović M., Kirćanski N., "Computer-Oriented Method for Linearization of Dynamic Models of Active Spatial Mechanisms", Journal of Mechanisms and Machine Theory, Vol. 16, No. 2, 1981.

[38] Vukobratović M., Kirćanski N., "Computer Method for Dynamic Model Linearization of Active Spatial Mechanisms", Journal "Mashinovedenie" ANUSSR, No. 6, 1982.

[39] Vukobratović M., Kirćanski N., "An Engineering Concept of Adaptive Control for Manipulation Robots via Parametric Sensitivity Analysis", Bulletin T. LXXXI de l'Académie Serbe des Sciences et des Arts, Classe des Sciences Techniques No. 20, 1982.

[40] Vukobratović M., Kirćanski N., "Computer Assisted Sensitivity Model Generation in Manipulation Robots Dynamics", Journal of Mechanism and Machine Theory, No. 1, 1984.

[41] Vukobratović M., Stokić D., Control of Manipulation Robots, Springer-Verlag, 1982.

[42] Kane T.R., "Dynamics of Nonholonomic Systems", ASME Journal of Applied Mechanics, Vol. 28, pp. 574-578, 1961.

[43] Kane T.R., Likins P., Kinematics of Rigid Bodies in Spaceflight, Applied Mechanics, Stanford University, Rep. 204, pp. 26-119, 1971.

[44] Huston R.L., Passerello C.E., "On the Dynamics of Chain Systems", ASME Paper No. 74-WA/Aut 11, 1974.

[45] Huston R.L., Posserello C.E., Harlow M.W., "Dynamics of Multirigid-Body Systems", ASME Journal of Applied Mechanics, Vol. 45, pp. 889-894, 1978.

[46] Whittaker E.T., Analytical Dynamics, London, Cambridge, 1937.

[47] Popov E.P., "Control of Robots-Manipulators", Journal "Teknitcheskaya kibernetika ANUSSR, No. 6, Moscow, 1974.

[48] Vukobratović M., Potkonjak V., "Two New Methods for Computer Forming of Dynamic Equations of Active Mechanisms", Journal of Mechanism and Machine Theory, Vol. 14, No. 3, 1979.

[49] Bourne S.R., Horton J.R., "The Design of the Cambridge Algebra System", Proceedings of the Second Symposium on Symbolic and Algebraic Manipulation, New York, Association for Computing Machinery, pp. 134-143, 1971.

[50] Manove M., Bloom S., Engleman C., "Rotational Functions in MATHLAB", Symbol Manipulation Langages and Techniques, Amsterdam, North-Holland, pp. 86-102, 1966.

[51] Martin W., Fateman R., "The MACSYMA Systems", Proceedings of the Second Symposium on Symbolic and Algebraic Manipulation, New York, Association for Computing Machinery, pp. 59-75, 1971.

[52] Hyland E., Chapman C., "Symbolic Algebra in Theoretical Seismology", Proc. of the Second Symposium on Symbolic and Algebraic Manipulation, New York, Assiciation for Computing Machinery, pp. 352-364, 1971.

[53] Dillon S., Computer Assisted Equation Generation in Linkage Dynamics, Ph. D. Thesis, Ohio State University, 1973.

[54] Aldon M.J., Liègeois A., "Génération et Programmation Automatiques des Equations de Lagrange des Robots et Manipulateurs", Rapport de Recherche, INRIA, 1981.

[55] Luh J.Y.S., Lin C.S., "Scheduling of Parallel Computation for Computer-Controlled Mechanical Manipulators", IEEE Trans. on SMC, Vol. 12, No. 2, pp. 214-234, 1982.

[56] Vukobratović M., Kirćanski N., "New Method for Real-Time Generation of Manipulator Dynamic Models" Proc. of 1st Yugoslav-Soviet Symp. on Applied Robotics, Moscow, 1983.

[57] Kirćanski N., Contribution to Dynamics and Control of Manipulation Robots, Ph.D. Thesis, Electrical Engineering Faculty, Belgrade, 1984.

[58] Vukobratović M., Kirćanski N., "Computer Aided Procedure of Forming of Robot Motion Equations in Analytical Forms", Proc. of VI IFToMM Congres, New Delhi, 1983.

[59] Kirćanski M., Contribution to the Synthesis of Manipulation Robots Control, MSc Thesis, Electrical Engineering Faculty, Belgrade, 1980.

[60] Vukobratović M., Kirćanski M., "A Dynamic Approach to Nominal Trajectory Synthesis for Redundant Manipulators", IEEE Trans. on SMC, Vol. 14, No. 4, 1984.

[61] Vukboratović M., Kirćanski N., "A Method for Computer-Aided Construction of Analytical Models of Robotic Manipulators", First IEEE Conf. on Robotics, Atlanta, 1984.

[62] Vukobratović M., Kirćanski N., "Computer Assisted Generation of Robot Dynamic Models in Analytical Form", Acta Applicandea Mathematicae, International Journal of Applying Mathematics and Mathematical Applications, Vol. 2, No. 2, 1984.

[63] Vukobratović M., Kirćanski N., "Implementation of High Efficient Analytical Robots Models on Microcomputers", V International Symposium on Theory and Practice of Robots and Manipulators, Udine, 1984.

[64] Vukobratović M., Kirćanski M., "One Method for Simplified Manipulator Model Construction and its Application in Quazioptimal Trajectory Synthesis", Journal of Mechanism and Machine Theory, Vol. 17, No. 6, pp. 369-378, 1982.

[65]   Leskov A.G. Medvedov V.S., "Analysis of Dynamics and Control synthesis of Manipulation Robots", Teknitcheskaya Kibernetika, ANUSSR, No. 6, Moscow, 1974.

[66]   Vereschagin A.F., "Computer-Aided Modelling Method of Manipulator Dynamics", Teknitcheskaya Kibernetika, ANUSSR, No. 6, Moscow 1974.

[67]   Krutko P.D., Lakota N.A., "Control Synthesis of Manipulation Robots Motion Based on Inverse Dynamics Solution", Teknitcheskaya Kibernetika, ANUSSR, No. 1, Moscow, 1981.

[68]   Vukobratović M., Li Shi-Gang, "One Efficient Procedure for Generating Dynamic Manipulator Models", Robotica (in press).

[69]   Paul R.P., Robot Manipulators: Mathematics, Programming, and Control, The MIT Press, Cambridge, Massachusetts, 1981.

M. Vukobratović, V. Potkonjak

Dynamics of Manipulation Robots

Theory and Application

1982. 149 figures. XIII, 303 pages (Communications and Control Engineering Series, Scientific Fundamentals of Robotics 1)
ISBN 3-540-11628-1

Contents: General Remarks about Robot and Manipulator Dynamics. – Computer-Aided Methods for Setting and Solving Mathematical Models of Active Mechanisms in Robotics. – Simulation of Manipulator Dynamics and Adjusting to Functional Movements. – Dynamics of Manipulators with Elastic Segments. – Dynamical Method for the Evaluation and Choice of Industrial Manipulators. – Subject Index.

In this monograph, complete mathematical models of the dynamics of open active mechanisms as applied to robotics are treated and presented for the first time.
It also presents the first parallel survey of computer-oriented methods for automatic construction of the dynamical equations of manipulation robots. These are based on a) general theorems of mechanics, b) second-order Lagrange's equations and c) Appel-Gibbs functions. Adjustment blocks are considered, thus enabling the dynamics of spatial mechanisms to be applied to concrete manipulation tasks. The dynamical nominal states (functional movements) of manipulation systems are synthesized and the elastic properties of the manipulation systems are considered.

Springer-Verlag
Berlin
Heidelberg
New York
Tokyo

M. Vukobratović, D. Stokić

Control of Manipulation Robots

Theory and Application

1982. 111 figures. XIII, 363 pages. (Communications and Control Engineering Series, Scientific Fundamentals of Robotics 2)
ISBN 3-540-11629-X

Contents: General Principles of Control Synthesis of Robots and Manipulators. – Dynamics of Manipulators. – Synthesis of Manipulation Control. – Appendix 2A: Suboptimality of Global Control. Appendix 2B: Analysis of "Distribution of the Model" Between Subsystems and Coupling. Appendix 2C: Stability Analysis of System with Decentralized Regulator and Observer. – Control Synthesis for Typical Manipulations Tasks. Appendix 3A: Simulation Algorithm for Assembly. – Subject Index.

Control synthesis principles and control algorithms of manipulation robots based on their exact dynamic mathematical models are presented in this volume for the first time in monograph form.

Springer-Verlag
Berlin
Heidelberg
New York
Tokyo